Rheinwerk Computing

The Rheinwerk Computing series offers new and established professionals comprehensive guidance to enrich their skillsets and enhance their career prospects. Our publications are written by the leading experts in their fields. Each book is detailed and hands-on to help readers develop essential, practical skills that they can apply to their daily work.

Explore more of the Rheinwerk Computing library!

Thomas Theis

Getting Started with Python

2024, 437 pages, paperback and e-book
www.rheinwerk-computing.com/5876

Philip Ackermann

JavaScript: The Comprehensive Guide

2022, 982 pages, paperback and e-book
www.rheinwerk-computing.com/5554

Philip Ackermann

Full Stack Web Development: The Comprehensive Guide

2023, 740 pages, paperback and e-book
www.rheinwerk-computing.com/5704

Michael Kofler

Scripting: Automation with Bash, PowerShell, and Python

2024, 470 pages, paperback and e-book
www.rheinwerk-computing.com/5851

Sebastian Springer

Node.js: The Comprehensive Guide

2022, 834 pages, paperback and e-book
www.rheinwerk-computing.com/5556

www.rheinwerk-computing.com

Thomas Theis

Getting Started with JavaScript

Editor Megan Fuerst
Acquisitions Editor Hareem Shafi
German Edition Editor Anne Scheibe
Translation Winema Language Services, Inc.
Copyeditor Doug McNair
Cover Design Mai Loan Nguyen Duy
Photo Credit Mai Loan Nguyen Duy
Layout Design Vera Brauner
Production Kelly O'Callaghan
Typesetting SatzPro, Germany
Printed and bound in the United States of America, on paper from sustainable sources

ISBN 978-1-4932-2583-5
© 2025 by Rheinwerk Publishing, Inc., Boston (MA)
1st edition 2025
5th German edition published 2024 by Rheinwerk Verlag, Bonn, Germany

Library of Congress Cataloging-in-Publication Control Number: 2024043360

Contents at a Glance

Contents

5 The Document Object Model

6 Using Standard Objects

8   Design Using Cascading Style Sheets   283

9   Two-Dimensional Graphics and Animations Using SVG   311

10   Three-Dimensional Graphics and Animations Using Three.js

11   jQuery

Materials for This Book

The following resources are available for you to download from this book's webpage:

- All sample programs
- All exercises with solutions
- Bonus chapters

Go to *www.rheinwerk-computing.com/5875* and scroll down to the **Product supplements**. You will see downloadable files along with brief descriptions of their contents. Click the **Download** button to start the download process. Depending on the size of the file (and your internet connection), you may need some time for the download to complete.

Chapter 1

Introduction

What is JavaScript, what can and can't I do with it, and how do I integrate it into my website? This chapter provides answers to initial questions like these.

JavaScript is an object-oriented programming language that was first developed in 1995. It's constantly updated on the basis of the ECMAScript standard, and a new version of the standard has been issued every year since 2015. The contents of this book are based on ECMAScript 2024.

The JavaScript language was designed for use in internet browsers. It's now also used beyond websites, but JavaScript's use in browsers is the topic of this book. Despite their similarities in individual aspects, JavaScript and Java—another widely used programming language—must be clearly distinguished from each other.

JavaScript is an *interpreted language*, which means JavaScript programs are translated and executed line by line.

JavaScript also offers many elements that may be familiar to you from other programming languages, such as loops for a quick repetition of program parts, branches for the different handling of different situations, and functions for breaking down a program into manageable components. You also have a variety of objects at your disposal. Using the *Document Object Model* (DOM), you can access all elements of your webpages in JavaScript so that you can change them dynamically.

1.1   What Can JavaScript Do?

The programs are available within internet pages, and they contain JavaScript code along with Hypertext Markup Language (HTML) code. It's important to note here that JavaScript provides additional possibilities and aids that HTML doesn't have.

After calling up a website with JavaScript code, the code is executed in the browser and can dynamically change the content of the website. This happens either immediately after loading or after the occurrence of an event such as your pressing a button when using the program. JavaScript enables the design of complex applications with a user interface.

During development, however, you should make sure that your code doesn't restrict the use of a website. For example, you should not write a JavaScript program that prevents you from leaving a website, because such a website won't be visited a second time.

Forms play the following important roles in connection with JavaScript:

- They are used to transmit data to a web server. Before sending, the contents of the formats can be checked for validity using JavaScript, which avoids unnecessary network traffic.

- They enable interactions similar to what you're used to with other computer applications. You can make entries and trigger processing, and the program will return the results.

To simplify the development of your programs, all you need is a text editor that can highlight the HTML markers and JavaScript keywords for clarity. The Notepad++ editor is one of many editors that can do this.

1.2   What Can JavaScript Not Do?

JavaScript can't switch itself on. There may be individual cases where JavaScript has been switched off in the browser, but the proportion of websites that can no longer be used in full is quite high. However, we can at least recognize whether JavaScript has been switched off and send a message if this is the case (Section 1.11).

JavaScript also can't save anything on the web server (without additions). JavaScript programs are executed in the browser and not on the web server from which they are loaded, so it's not possible to save data on the web server.

JavaScript can only store a small amount of data in the browser during use. It can't cause any damage there.

1.3   Browsers and Mobile Browsers

Internet pages, with or without JavaScript, are received by different browsers on different operating systems on different end devices and implemented during use.

Various browser statistics are available on the internet, for example at *https:// www.w3schools.com/browsers/*, and they can be used to determine which browsers are used and how often. For years, the *Google Chrome* browser has had the highest user share, and it therefore serves as the most important (but not the only) reference for the sample programs in this book.

Some JavaScript statements may be formulated differently for certain browsers, but I recommend always using the standard form so that the statement is suitable for as many browsers as possible. This principle also applies to the sample programs in this book.

The proportion of mobile devices with mobile browsers tailored to them is increasing all the time. Mobile devices provide some additional options such as receivers or sensors for geolocation data, plus the position and acceleration of the mobile device. The data determined in the process can be further processed by JavaScript (see Chapter 15, Section 15.3).

1.4   ECMAScript

JavaScript is based on *ECMAScript*. You can apply object-oriented programming in JavaScript in the traditional way, on the basis of prototypes and constructor functions. Since the introduction of the ECMAScript 2015 standard, there has been a new standard notation that uses classes, constructors, properties, and methods, as do other object-oriented languages. In this book, we use only this kind of notation, which is also implemented by all modern browsers. More on this will follow in detail in Chapter 3.

The ECMAScript standard has been updated annually since 2015 and the ECMAScript 2024 standard has been completed since July 2024. You can find an overview of many standardized elements and their implementation in individual browsers on the following website: *https://compat-table.github.io/compat-table/es2016plus*.

Many elements of the newer standards are also useful for beginners and can be found in the relevant sections of this book. I will introduce each of them separately.

1.5   Structure of This Book

First of all, a note for my own behalf: I would like to thank the team at Rheinwerk Germany, especially Anne Scheibe, and Rheinwerk US, especially Megan Fuerst, for their help in creating this book in its German and English editions.

I present each topic in this book with a brief description of the theory behind it; a meaningful screenshot; a complete, executable sample program; and a detailed practical explanation. The screenshots were taken in the Google Chrome browser, either on a PC with Windows 11 or on an Android smartphone. This way, you get a quick introduction to each topic, and you are not forced to place individual lines of code in a suitable context first to consider their effect. You'll find all sample programs in the downloadable materials for this book, which you can access at *www.rheinwerk-computing.com/5875*.

You'll also find references to exercises in this book, and you can find those tasks in the bonus chapters that are also included in the downloadable materials. The exercises allow you to test your knowledge, and you'll find a solution to each exercise in the downloadable materials.

The contents of this book build on each other in small, clear steps. This has the advantage that the prerequisites for each topic are clarified for you in advance, but it has the disadvantage that you need to actually read the book from cover to cover. If you simply open it at any point, you can't assume that all the details will be explained there, as they may have been included in an earlier section.

This chapter serves as the introduction to JavaScript, and the basic principles of JavaScript programming follow in Chapter 2, where you'll find that JavaScript is similar to many other programming languages.

Objects also play a major role in JavaScript, and in Chapter 3, you'll create your own objects and learn about their structure. In Chapter 6, I will explain many predefined JavaScript objects.

For interaction during use, we use events and their consequences, especially in connection with forms. See Chapter 4 for an explanation of this. Knowledge of the structure of a web document according to the DOM enables you to access and change any part of the document, and Chapter 5 goes over this in detail.

Ajax technology (see Chapter 7) allows you to exchange individual parts of a document without having to completely reload a page. *Cascading Style Sheets* (CSS) offer a wide range of options for formatting and positioning elements of an HTML document. These are expanded dynamically using JavaScript, right up to animation, as you'll see in Chapter 8.

Dynamic, two-dimensional vector graphics can be created using JavaScript and the *Scalable Vector Graphics* (SVG) standard, which is described in Chapter 9. The *Three.js* JavaScript library provides the option to develop three-dimensional graphics and animations, as I explain in Chapter 10.

The widely used *jQuery* library enables browser-independent, convenient access to many JavaScript elements, as I explain in Chapter 11, and the *Onsen UI* library is used specifically for programming mobile devices, as I explain in Chapter 12. Mathematical expressions can be displayed in your documents using *MathML* and the *MathJax* JavaScript library, as you'll see in Chapter 13.

In Chapter 14, I refer to a series of larger, extensively commented sample projects that show the interactions among many elements. Finally, in Chapter 15, I explain how to access media and sensors and create drawings.

1.6   First Example with HTML and CSS

Only a little knowledge of HTML and CSS is required to understand the examples in this book. The most important elements are explained using the following initial sample program.

1.6.1   Output of the Program

In Figure 1.1 and Figure 1.2, you can see the results of the program in the browser.

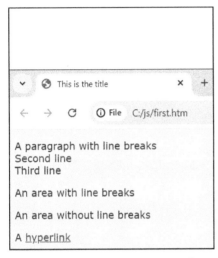

Figure 1.1 First HTML Document in the Browser, Upper Part

Figure 1.2 First HTML Document in the Browser, Lower Part

1.6.2   HTML File

The sample program contains some basic elements of an HTML document.

Here's the HTML code:

```
<!DOCTYPE html><html lang="en-us">
<head>
    <meta charset="utf-8">
    <title>This is the title</title>
    <link rel="stylesheet" href="js5.css">
</head>
<body>
    <p>
        A paragraph with line breaks
        Second line<br>Third line
    </p>
```

```
An area with line breaks
<p>An area <span>without</span> line breaks</p>

<p>A <a href="embed.htm">hyperlink</a></p>
<p>An image:<br><img src="im_paradise.jpg" alt="Paradise"></p>

<p>A list:</p>
<ul>
   <li>First entry</li>
   <li>Second entry</li>
</ul>

<p>A table:</p>
<table>
   <tr>
      <td>Cell A</td>
      <td>Cell B</td>
   </tr>
   <tr>
      <td>Cell C</td>
      <td>Cell D</td>
   </tr>
</table>
</body>
</html>
```

Listing 1.1 first.htm File

You use `<!DOCTYPE html>` to specify that the document is an HTML document. The more you adhere to the standardized definitions of HTML documents, the higher the probability that the page will be displayed correctly in all browsers.

An HTML document consists of *markers* (also called *tags*) and text. Most markers form a *container* that has a start marker and an end marker. Start markers can have attributes with values, and the latter are placed in double quotation marks by default.

The entire document is in the `html` container from the start marker `<html>` to the end marker `</html>`. The start marker `html` has the attribute `lang` with the value `en-us`, which indicates that the text of the document is written in US English. The `html` container contains a `head` container with information about the document and a `body` container with the actual document contents.

In the `head` container, you'll first find a `title` container, which provides the content for the title bar of the browser. Metadata about the document can also be found here. In this example, you can see that this is an HTML document that uses the widely used 8-bit UCS Transformation Format (UTF-8) character set (Section 1.6.3), which contains many special characters, including German umlauts, for example.

You can use the `link` marker and the `rel` and `href` attributes to include an external CSS file for formatting the document. In this example, the file is *js5.css*, which is located in the same directory as the *first.htm* file. The CSS file is explained in more detail in Section 1.6.4.

Paragraphs are stored in `p` containers, and a single line break within a paragraph is created using the `<br>` marker. You can place certain areas that you want to format differently in a `div` container as well as in a `span` container. In addition, a line break is created before and after a `div` container.

A clickable hyperlink to another document is located in an `a` container with the `href` attribute. An image can be integrated using the `img` marker and the `src` attribute. The `alt` attribute is required for validation; it contains explanatory text in case the image file can't be loaded.

An unnumbered list is in a `ul` container, and the individual list entries are placed in `li` containers.

Tables can be created using a `table` container. There are individual rows within the table; each of these is created using a `tr` container. Within a row, in turn, there are individual cells, each of which is formed by a `td` container.

You can create the *first.htm* file using the Notepad++ editor and save it in the *C:/js* directory (the latter is my preference). To display an *htm* file (or an *html file*) in your default browser, you need to open Windows Explorer and double-click on the *htm* file.

If other HTML markers are used, they are explained in the appropriate place.

1.6.3   UTF-8 Encoding

The UTF-8 encoding is used in all HTML documents in this book. *UTF-8* is an abbreviation for the *8-bit UCS Transformation Format*, and *UCS* stands for *Universal Coded Character Set*. UTF-8 is the most widely used encoding for Unicode characters.

It's important that the encoding specified in the `head` container matches the encoding of the file. You can change the encoding of a file to UTF-8 in the Notepad++ editor as follows: **Menu Encoding • Convert to UTF-8**. Once you have done that, the UTF-8 encoding will be shown as selected in this menu.

You can also automatically select the encoding in the Notepad++ editor as follows for all new files you create: **Menu Settings • Preferences • New Document • Encoding • UTF-8**. Then, click the **Close** button.

1.6.4   Responsive Web Design

In this first example, the external CSS file *js5.css* is used to format the document. It uses a *media query* to create a responsive web design in a simplified manner. This achieves the following goals:

- The documents are formatted in a uniform way.
- If required, the formatting can be changed quickly and uniformly for all documents.
- The examples can be used not only on a PC or laptop but also on a mobile device.

Here's the code used in the CSS file:

```css
body {font-family:Verdana; font-size:11pt; color:#202020;
      background-color:#f8f8f8;}
td   {font-size:11pt; background-color:#e0e0e0; padding:5px;}

@media only screen and (max-width: 992px)
{
   @media only screen and (orientation:landscape)
   {
      body                 { font-size:20pt; }
      td                   { font-size:20pt; }
      img                  { width:240px; height:180px; }
      input                { font-size:20pt; }
      select               { font-size:20pt; }
      input[type=radio]    { width:30px; height:30px; }
      input[type=checkbox] { width:30px; height:30px; }
      input[type=color]    { width:250px; height:30px; }
      input[type=range]    { width:250px; height:30px; }
      textarea             { font-size:20pt; height:80px; }
   }
   @media only screen and (orientation:portrait)
   {
      body                 { font-size:32pt; }
      td                   { font-size:36pt; }
      img                  { width:320px; height:240px; }
      input                { font-size:32pt; }
      select               { font-size:32pt; }
      input[type=radio]    { width:45px; height:45px; }
      input[type=checkbox] { width:45px; height:45px; }
      input[type=color]    { width:250px; height:45px; }
      input[type=range]    { width:250px; height:45px; }
      textarea             { font-size:32pt; height:140px; }
   }
}
```

Listing 1.2 js5.css File

What follows is just a brief explanation of CSS. In Chapter 8, you'll find more information on this topic.

A CSS specification can apply to an HTML marking to which the formatting refers. Curly brackets contain one or more formatting options, which in turn consist of a CSS property and a value for this property, separated by a colon and terminated by a semicolon.

In this example, a number of formatting settings are made for the body marker (i.e., the content of the document):

- The Verdana font is used via font-family.
- The font size of 11 points is determined via font-size.
- The #202020 font color is specified by color.
- The background color #f8f8f8 is determined by background-color.

You can specify colors by using RGB values. The # character is followed by two hexadecimal digits for the red, green, and blue color components. The color #202020 corresponds to a dark gray, while #f8f8f8 corresponds to a very light gray.

The font size within table cells (marking td) is separately set to 11 points. The background color of table cells is set to a light gray by using #e0e0e0, and this makes the cells slightly darker than the background of the document. By using the padding specification, you can adjust the inner distance between an element and its surrounding element. Here, a distance of 5 pixels is set around the text content of a table cell to the edge of the table cell.

This is where the CSS information for use on a PC or laptop ends.

You can use a media query to respond to the properties of the devices on which you view the documents. The CSS specification @media only screen and (max-width: 992px) causes the CSS specifications inside the subsequent curly brackets to do the following:

- Reference only an output on a screen, in contrast to an output on a printer.
- Apply to devices with a maximum output width of 992 pixels, such as mobile devices like tablets and smartphones.

The orientation is then used to differentiate whether the tablet or smartphone is currently in landscape mode or portrait mode.

The font size in the document is increased to 20 points in landscape format and 32 points in portrait format. Table cells are enlarged to 20 points or 36 points, respectively.

Many images in my sample programs have a uniform size of 160 × 120 pixels for simplification. In landscape format, they are enlarged to 240 × 180 pixels, and in portrait format, they are enlarged to 320 × 240 pixels.

The following HTML elements are used to design forms. You will get to know them in Chapter 4. Their size is also adjusted depending on the media.

- General input, single-line text field (`input`)
- Selection using a menu (`select`)
- Selection using radio buttons (`input type=radio`)
- Checkbox (`input type=checkbox`)
- Selection of a color (`input type=color`)
- Setting a number in a range (`input type=range`)
- Multiline text field (`textarea`)

1.7   Some Special Characters

You use the following program to output some special characters. You can also enter some special characters directly using the keyboard, and all of them can be output using so-called *entities*.

Here's the program:

```
...
<body>
    <p>Some special characters on the keyboard: > & $ @<br>
        also the entity: &gt; &amp; &dollar; &#64;<br>
        Some other special characters: &copy; &reg; &permil;
            &frac14; &frac12; &frac34; &sup2; &sup3; &micro; &pi;<br>
        The &lt; character is only validated as an entity</p>
</body></html>
```

Listing 1.3 special_characters.htm File

> **Note**
>
> In this example and many of the following examples, the beginning of the document is omitted to save space. It's only printed if it contains additional information, and the end of the document is presented in a more compact form.

You can insert the special characters in the first line directly into your document using the ⟩ (greater than), ⟨&⟩ (ampersand), ⟨$⟩ (dollar), and ⟨@⟩ (at) keys.

There are entities available for a large number of special characters. An entity begins with the & character and ends with a semicolon:

- In the second line you can see the entities for >, &, $, and @: >, &, $, and @, respectively.
- The third line contains the entities for © (copyright), ® (registered trademark), ‰ (per thousand), ¼ (one quarter), ½ (one half), ¾ (three quarters), [2] (to the power of 2), [3] (to the power of 3), µ (micro), and π (Pi): ©, ®, ‰, ¼, ½, ¾, ², ³, µ, and π, respectively.
- You should output the < (less than) character in the fourth line using the < entity; otherwise, the associated document won't be validated as an HTML document.

In Figure 1.3, you can see the document in the browser.

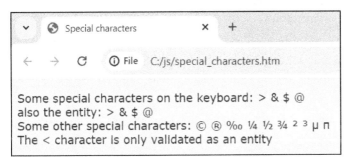

Figure 1.3 Some Special Characters

1.8 JavaScript in the Document

Now, it's finally time to get started with the first JavaScript program. First, I want you to take a look at the following code:

```
...
<body>
    <p>Text in HTML</p>
    <script>
        document.write("<p>Text in JavaScript</p>");
    </script>
    <p>Text in HTML</p>
    <script>
        document.write("<p>JavaScript, double quotation marks</p>");
        document.write('<p>JavaScript, single quotation marks</p>');

        document.write("<p>Single quotation marks<span style='font-weight:bold;'>"
            + " inside</span> double quotation marks</p>");
```

```
    document.write('<p>Double quotation marks <span style="font-weight:bold;">'
        + ' inside</span> single quotation marks</p>');
    </script>
</body></html>
```

Listing 1.4 embed.htm File

You can embed JavaScript in any number of places in the head or body of an HTML document. In each case, a script container is required. This container begins with `<script>` and ends with `</script>`.

The container contains JavaScript statements that are executed one after the other. They are concluded with a semicolon.

The `document.write()` call outputs a character string. The *dot* operator is placed between the `document` object and the `write()` method, and this calls the `write()` method for the `document` object. I explain the term *object* in detail in Chapter 3.

By default, character strings are enclosed in double quotation marks. The character strings can contain both text and HTML markers, but you can also enclose strings in single quotation marks.

The same applies to the values of attributes, as you can recognize by the value of the `style` attribute. In HTML, they are written in double quotation marks by default, but you can also enclose attribute values in single quotation marks.

To avoid errors when outputting attribute values that are within a character string for JavaScript, different quotation marks must be combined. These are either single quotation marks within double quotation marks or vice versa.

I have used the `font-weight` CSS property with the `bold` value to set part of the text in bold.

Useful Information

- Observe the correct notation of the statements when programming. Unlike in HTML, browsers don't forgive errors in JavaScript.

- JavaScript distinguishes between uppercase and lowercase letters. You'll not be successful with the `document.Write(...)` statement, as the `Write()` method with an uppercase `W` doesn't exist.

- You can also write multiple statements in one line. The main thing to remember is that there's a semicolon at the end of each statement.

In Figure 1.4, you can see various parts of the document, some of which originate from HTML and some from JavaScript.

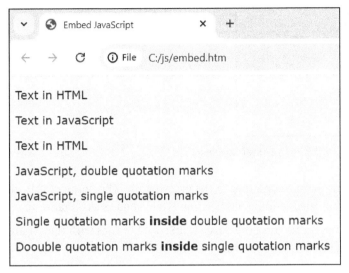

Figure 1.4 JavaScript within a File

1.9   JavaScript from an External File

You can save program parts that you want to use in several JavaScript programs in an external file. You can easily access the code of such an external file by integrating the file into your program.

Here's an example:

```
...
<body>
   <script src="external_file.js"></script>
   <script>
      document.write("<p>This comes from external.htm</p>");
   </script>
</body></html>
```

Listing 1.5 external.htm File

The first `script` container is empty, but the `src` attribute is noted with the `external_file.js` value. This integrates the code from the relevant file into the *external.htm* file. The *external_file.js* file only contains the following code:

```
document.write("<p>This is from external_file.js</p>");
```

Listing 1.6 external_file.js File

In Figure 1.5, you can see the two paragraphs that are generated from the merged program using the `document.write()` method.

31

Figure 1.5 Additional JavaScript from an External File

Please note that there's no `script` container in the external file. The name of this file can have any extension, but the *js* ending has become established as a convention.

The jQuery library (see Chapter 11) and other large JavaScript libraries with their many useful functions are integrated into applications in this way.

1.10   Comments

Comments are used to describe the individual parts of your programs, and they make it easier for you and others to understand those programs. Let's look at an example:

```
...
<body>
    <!-- This is a comment
        in the HTML section -->
    <p>A paragraph from the HTML section</p>
    <script>
        /* This is a comment across multiple lines
           in the JavaScript section */
        document.write("<p>A paragraph from the JS section</p>");
        // A short comment, only to the end of the line
    </script>
</body></html>
```

Listing 1.7 comment.htm File

In the example, you can see three different types of comments:

- A comment in the HTML section can run across one or more lines. It's located between the character strings `<!--` and `-->`.

- In the JavaScript section, a comment that spans one or more lines is noted between the /* and */ strings.

- If you want to write a short comment in the JavaScript section (for example, after a statement), the // string is suitable. Such a comment is only one line in length.

The content of the comments is not displayed in the browser, as you can see in Figure 1.6. However, anyone can view the source code of a page in their browser if required and thus also the comments.

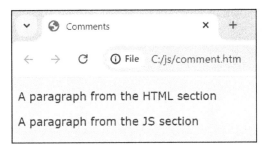

Figure 1.6 Comments Are Not Visible in Browser

> **Note**
> Long after you create a program, you'll often want to look at or expand it. Then, you'll be grateful for every line of commentary you find in it. For the same reasons, it's also highly recommended that you write clear, easy-to-read programs.

1.11   No JavaScript Is Possible

As already mentioned in Section 1.2, there may be individual cases where JavaScript has been switched off in the browser. Since JavaScript can't switch itself on, what can you do?

You can recognize whether it's switched on or not. If it's not switched on, you can either offer a simple version of the page in pure HTML or indicate that the use of the page in question requires JavaScript to be switched on.

Here's an example:

```
...
<body>
   <script>
      document.write("<p>JavaScript is running here</p>");
   </script>
   <noscript>
      <p>JavaScript is not running here<br>
         Please switch it on</p>
   </noscript>
</body></html>
```

Listing 1.8 no_script.htm File

Within the noscript container, you can note down text and HTML markers in the event that JavaScript is switched off.

If JavaScript is enabled, only the statements from the script container will be executed, and the page will then look like the one shown in Figure 1.7.

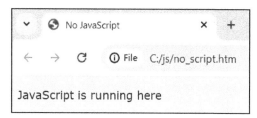

Figure 1.7 JavaScript Is Enabled

You can deactivate JavaScript once for testing purposes, and the necessary procedure is explained in the following text using the example of the Google Chrome browser. Call the menu using the three dots at the top right, select the **Settings** menu item, and enter the term "JavaScript" in the search window. You'll find the **Site settings** in the search results. There, you'll find the **JavaScript** area, with the option **Sites can use JavaScript** by default. Select the other option **Don't allow sites to use JavaScript**.

Then, when you call a file that contains JavaScript, you'll see an icon on the far right of the browser address bar with the information that JavaScript has been blocked on this page. Click on the icon to open a dialog box. There, you can activate JavaScript *for this page*. The **Manage** button also takes you directly to the previously mentioned options for switching JavaScript on and off.

When JavaScript is switched off, the *no_script.htm* page looks like Figure 1.8.

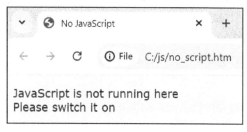

Figure 1.8 JavaScript Is Disabled

Chapter 2
Basic Principles of Programming

Variables, operators, branches, loops, and functions are the basic elements you need to get started with programming.

In this chapter, you'll learn the basics of programming in JavaScript: variables, operators, branches, loops, and functions. These elements form the basis of many programming languages and will make it easier for you to get started in other areas later on.

2.1 Storing Values

Texts or numbers often have to be saved within a program sequence, and they can result both from assignments during the development of the program and from entries made when using the program. These saved values are required again at a later point in the program.

This section deals with the storage of texts, numbers, and truth values using *variables*, which can be changeable or unchangeable. Unchangeable variables are also referred to as *constants*. In this section, you'll also get to know two dialog boxes for simple inputs and outputs.

Variables don't have a defined type in JavaScript. You can save a character string (i.e., a text) in a variable, and if the variable is changeable, you can save a number in the same variable later.

2.1.1 Formatting Strings

Texts consist of individual characters, and in their entirety, they are called *character strings*. Variables are needed for storing values. The following program contains some examples:

```
...
<body><p>
   <script>
      let helloText;
      helloText = "Hello world";
      document.write("Text: " + helloText + "<br>");
      helloText = "Good morning";
```

```
        document.write("Text: " + helloText + "<br>");

        let text2 = "Declaration and value";
        document.write("Text: " + text2 + "<br>");

        let text3 = 'With single quotation marks';
        document.write("Text: " + text3 + "<br>");

        let text4;
        document.write("Text: " + text4 + "<br>");

        // document.write("Text: " + text5 + "<br>");
        // text5 = "No declaration";
        // document.write("Text: " + text5 + "<br>");

        const text6 = "Cannot be changed";
        document.write("Text: " + text6 + "<br>");
        // text6 = "Not allowed";
        document.write("End");
    </script></p>
</body></html>
```

Listing 2.1 strings.htm File

A *declaration* is used to create storage space for the value of a variable. This space can be reached using a name, and the rules for assigning names can be found in Section 2.1.2.

At the start of the program, the JavaScript keyword let is used to declare a variable named helloText.

After the declaration, the helloText variable is first assigned the "Hello world" value, and a little later, it is assigned the "Good morning" value. The specific value of the variable is output on the screen using document.write(), as you can see in Figure 2.1.

An output can consist of several parts. This can involve a string in double quotation marks, which can also include HTML tags. However, it can also be a variable whose current value is output. The individual parts are combined into a longer string using the + operator.

Using the text2 variable, the declaration and assignment are combined within one statement.

The value of a string can also be written between single quotation marks, as you can see in the text3 variable.

The text4 variable is declared, but it doesn't receive a value in the program code. It therefore has the value undefined.

Read access to a variable that hasn't been declared and hasn't yet been assigned a value is not permitted. The first program line with the text5 variable would lead to an error and a program abort, which is why the line has been commented out.

Although write access to a variable that hasn't been declared is permitted, it's not recommended because the variable can be changed more easily by mistake and is more difficult to control. For this reason, the other program lines with the text5 variable have been commented out as well.

If you want to mark a variable as unchangeable and at the same time ensure that its value can't be changed in the further course of the program, you should use the const keyword instead of let. The variable text6 is such an unchangeable variable (i.e., a constant).

Assigning a new value to the constant would lead to an error and a program abort, so this line has also been commented out.

You can remove the commenting from one of the above-mentioned incorrect program lines once for test purposes. The output of the word End shows you whether the program has reached the end or has been terminated prematurely.

The output of the program is shown in Figure 2.1.

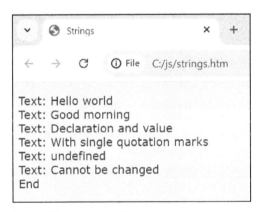

Figure 2.1 Output of Strings

Note

Variables can also be declared using the var keyword, but these variables can be changed more easily by mistake and are more difficult to control. This is a particular disadvantage with longer programs that consist of a large number of program sections, so the use of these variables can result in program errors more quickly. Instead, you should always use the let and const keywords, as is done in all the examples in this book.

2.1.2   Naming Rules

Here are some rules for naming a variable:

- A name can consist of uppercase and lowercase letters as well as numbers and the underscore (_).
- There must not be a number at the beginning of the name.
- Special characters, such as the German umlauted letters ä, ö, ü, Ä, Ö, and Ü, the ß, or a space in the name, are not permitted.
- The name must not correspond to one of the JavaScript keywords. A list of keywords can be found in Appendix A, Section A.2.

Also, here are two common conventions:

- The name of a variable should be self-explanatory (i.e., it should tell you something about the content). If a family name is to be saved, you should also name the variable `familyName`.
- Longer names for variables are particularly self-explanatory. *Camel case* notation has become established: the name begins with a lowercase letter and each new word begins with a capital letter (e.g., `buttonStart`, `buttonStop`).

2.1.3   Input and Output of Strings

In this section, you'll learn about two useful methods:

- The `prompt()` method, which opens a dialog box and prompts you to input a text
- The `alert()` method, which opens a dialog box that displays a message

Both methods use strings and cause the program to stop in its course, after which, the program only starts running again after an entry has been made or the message has been confirmed. This gives you the opportunity to interact, request an input, or initiate the reading of an output.

Here's an example:

```
...
<body>
   <script>
      const input = prompt("Input:\nPlease make an entry:",
         "Good morning");
      alert("Your input:\n" + input + "\nThank you very much");
   </script>
</body></html>
```

Listing 2.2 strings_in_out.htm File

As soon as you start the program, the prompt shown in Figure 2.2 appears.

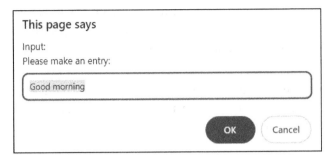

Figure 2.2 Input with prompt()

The prompt() method creates the prompt with the text that comes from the first string, and the \n control character creates a line break within a dialog box. The letter n stands for *new line*.

You can optionally specify a second string for prompt(), separated by a comma. This contains a possible default for the input field: in this case, the text Good morning. If you omit the default setting, the input field will be empty.

After you click the **OK** button, the current content of the input field gets returned and saved in the input variable.

The null value represents a *reference to nothing*. If you click the **Cancel** button in this dialog box, this value gets returned and saved as null. In Section 2.3.5, we determine whether the **OK** button or the **Cancel** button has been clicked.

If you change the input to This is my input, a message like the one shown in Figure 2.3 is displayed due to alert().

Figure 2.3 Output with alert()

The \n control character is also used in this dialog box to create a line break, and the output is combined into a longer string using the + operator.

> **Note**
>
> If a statement is very long, you can write it in several lines in the editor. You can implement a line break in many places within a statement, but not in the middle of a keyword and not within a string.

The alert() method is useful for troubleshooting. You can use it to display the values of variables at several points in the program, and if they don't match the expected values, you can narrow down the location of the cause of the error. You'll learn more about this topic in Section 2.5.

Both prompt() and alert() are methods of the window object. Your call should therefore actually read window.prompt() or window.alert(), but this is not necessary for the window object in particular. I explain the term *object* in detail in Chapter 3.

> **Note**
>
> You'll find the u_input exercise in bonus chapter 1, section 1.1, found in the downloadable materials for this book at *www.rheinwerk-computing.com/5875*.

2.1.4   Storing Numbers

Numbers are required for calculations and are also stored in changeable or unchangeable variables. These numbers can result from assignments during the development of the program as well as from inputs when using the program. In this context, it's important to note a few special features, as the following example shows:

```
...
<body><p>
    <script>
        let number;
        number = 42;
        const anotherNumber = number + 30.8;
        const smallNumber = -3.7e-3;
        const bigNumber = 5.2e6;
        const manyDigits = 3_530_755.383_725;

        document.write("<p>First number: " + number + "<br>");
        document.write("Another number: " + anotherNumber + "<br>");
        document.write("Addition incorrect: " + anotherNumber + 25 + "<br>");
        document.write("Addition correct: " + (anotherNumber + 25) + "<br>");
        document.write("Small number: " + smallNumber + "<br>");
        document.write("Big number: " + bigNumber + "<br>");
        document.write("Many digits: " + manyDigits);
    </script></p>
</body></html>
```

Listing 2.3 numbers.htm File

First, a changeable variable is declared using let, and a numerical value is assigned to this variable.

Next, a constant is declared using const, a calculation takes place on the right-hand side of the statement, and the result of the calculation is assigned to the constant. Please note that decimal places must be denoted by periods. I do the same in these explanations.

The + operator is placed between two numbers and therefore has an effect (i.e., two numbers are added) that's different from when it is placed between two texts (i.e., two texts are joined together).

In the program, two values follow in *exponential format*: $-3.7 \times 10^{-3} = -0.0037$ and $5.2 \times 10^6 = 5.2$ million. This format is particularly suitable for very small and very large numbers, as it saves you the trouble of entering many zeros (as well as the likely errors involved).

Finally, a number is assigned that has many digits both before and after the decimal point. ECMAScript 2021 introduces the _ (underscore) character as a *numeric separator*, which is an input and reading aid for numbers with many digits and is already used by all modern browsers. It makes sense to insert the underscore after every three digits.

The output of the program is shown in Figure 2.4.

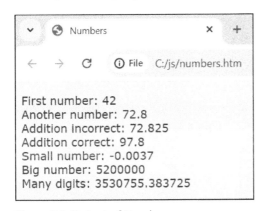

Figure 2.4 Output of Numbers

Let's look at the program code for the output using document.write(). In the first line, a character string is connected to a number using the + operator. This involves an automatic type conversion. The number 42 becomes the string "42", and in total, this results in the string "First number: 42". The same applies to the second line.

This procedure leads to an unexpected result in the third line. After the type conversion of the two numbers, the three strings "Addition incorrect: ", "72.8", and "25" become the long character string "Addition incorrect: 72.825". If the two numbers are to be added up first, then parentheses must be used, as shown in the next line. Only then will the correct result of the addition be displayed.

Finally, the small number, the large number, and the number with many digits are output.

2.1.5   Storing Truth Values

Truth values (also known as *Boolean values*) can also be stored in a (changeable or unchangeable) variable. There are two possible values: `true` and `false`. In this way, you can save the status of a *checkbox* in a form, for example. If the checkbox is selected, the `checked` property of the checkbox (see Chapter 4, Section 4.4) has the value `true`; if the checkbox is not selected, the `checked` property of the checkbox has the value `false`.

In this way, information is stored for which there are only two states. The options for states are *on* or *off*, 0 or 1, and *correct* or *incorrect*.

Here's an example:

```
...
<body>
   <script>
      let saved;
      saved = true;
      const done = false;
      document.write("<p>Saved: " + saved + "<br>");
      document.write("Done: " + done + "</p>");
   </script>
</body></html>
```

Listing 2.4 bool.htm File

The `saved` variable is assigned the `true` value, while the `done` constant is assigned the `false` value. Then, they are output, as shown in Figure 2.5.

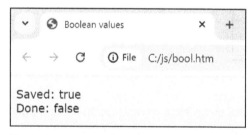

Figure 2.5 Output of Truth Values

2.2   Performing Complex Calculations

The following sections show you how to perform calculations with variables and numerical values. Your programs become interactive through the input of numbers and the output of results.

2.2.1 Calculation Operators

In addition to the + operator, the following operators are available for performing calculations: - for subtraction, * for multiplication, / for division, and % for modulo (the last of which calculates the remainder of an integer division). Since ECMAScript 2016, there's also the ** operator for exponentiation.

The following program does a little math:

```
...
<body><p>
   <script>
     let z;
     z = 2 + 4 - 2.5;       document.write("2 + 4 - 2.5 = " + z + "<br>");
     z = 2 + 4 * 3 - 6 / 2; document.write("2 + 4 * 3 - 6 / 2 = " + z + "<br>");
     z = (2 + 4) * (3 - 6) / 2;
                           document.write("(2 + 4) * (3 - 6) / 2 = " + z + "<br>");
     z = 13.5 % 5;          document.write("13.5 % 5 = " + z + "<br><br>");

     z = 6;                 document.write("z = " + z + "<br>");
     z = -z;                document.write("z = -z | z = " + z + "<br>");
     z = z * 5;             document.write("z = z * 5 | z = " + z + "<br><br>");

     z = 2.5 ** -3.5;       document.write("2.5 ** -3.5 = " + z + "<br>");
     z = (-2.5) ** -3;      document.write("(-2.5) ** -3 = " + z + "<br>");
     z = (-2) ** -3.5;      document.write("(-2) ** -3.5 = " + z + "<br>");
     z = 4 * 3 ** 2;        document.write("4 * 3 ** 2 = " + z);
   </script></p>
</body></html>
```

Listing 2.5 calculation.htm File

After each calculation, the calculation itself and the corresponding result are displayed for a better overview.

Of course, the PEMDAS rules of mathematics apply here. This means that multiplication and division have a higher priority than addition and subtraction and are therefore carried out first. If multiple operators have the same priority, the calculations are carried out from left to right. You must set parentheses to change priorities.

First, do the calculations in the first of the three blocks in Listing 2.5:

- The first calculation is performed from left to right and results in the value 3.5.

- In the second calculation, 4 * 3 and 6 / 2 are calculated first. Only then can an addition or subtraction operation follow. The result is 2 + 12 − 3 = 11.

- In the third calculation, the contents of the parentheses (2 + 4) and (3 – 6) are calculated first. Only then can the remaining calculations be carried out. The result is 6 * –3 / 2 = –9.

- If we divide 13.5 by 5, we get "2 with a remainder of 3.5". The modulo operator % determines the remainder, 3.5.

Then follow the calculations from the second block:

- As with numbers, you can precede variables with a minus sign. This multiplies the value of the variable by –1, so 6 becomes –6 and –6 becomes 6.

- A variable can appear on both sides of an assignment. In the last example, you can see that the old value of z (–6) is multiplied by 5 to result in the new value of z (–30).

The third block deals with the exponentiation operator **, which has an even higher priority than multiplication and division:

- The first calculation is $2.5^{-3.5}$. The value before the ** operator is called the *base*, and the value after the operator is the *exponent*. Both can have decimal places, and the exponent can be preceded by a negative sign.

- If there's a negative sign in front of the base, the value must be placed in parentheses. You can see this in the second example.

- The third example shows that the exponentiation of a negative number with a negative number that has decimal places is mathematically not permitted. This results in the NaN value, which stands for *not a number*.

- In the fourth example, due to the higher priority, the exponentiation is carried out first, followed by the multiplication. The result is $4 * 3^2 = 4 * 9 = 36$.

Figure 2.6 shows the results.

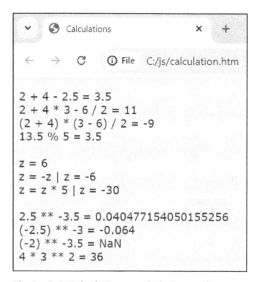

```
2 + 4 - 2.5 = 3.5
2 + 4 * 3 - 6 / 2 = 11
(2 + 4) * (3 - 6) / 2 = -9
13.5 % 5 = 3.5

z = 6
z = -z | z = -6
z = z * 5 | z = -30

2.5 ** -3.5 = 0.040477154050155256
(-2.5) ** -3 = -0.064
(-2) ** -3.5 = NaN
4 * 3 ** 2 = 36
```

Figure 2.6 Calculations and Their Results

2.2.2   Combined Assignment

You can shorten statements by using a *combined assignment*. The assignment operators +=, ++, and -- are frequently used, while the assignment operators -=, *=, /=, and %= are used less often. All of these operators are used in the following example:

```
...
<body>
  <script>
    let tx;
    tx = "This";         document.write("<p>" + tx + "<br>");
    tx = tx + " is a";   document.write(tx + "<br>");
    tx += " sentence.";  document.write(tx + "</p>");

    let z;
    z = 6;       document.write("<p>" + z + " ");
    z++;         document.write(z + " ");
    z--;         document.write(z + " ");
    z += 13;     document.write(z + " ");
    z -= 5;      document.write(z + " ");
    z *= 3;      document.write(z + " ");
    z /= 6;      document.write(z + " ");
    z %= 3;      document.write(z + "</p>");

    let a = 5, b = 5, c = 5, d;
    d = a++;     document.write("<p>" + d + " ");
    d = ++b;     document.write(d + " ");
    d = c++ + ++c + c++;   document.write(d + "</p>");
  </script>
</body></html>
```

Listing 2.6 assignment_combined.htm File

First, a string is assigned and extended twice. The assignment operator += is used for the second extension.

Then, a variable is assigned and changed several times. The ++ operator increases the value of a variable by 1, whereas the -- operator decreases it by 1. You would achieve the same result by using the z = z + 1 or z = z - 1 statements.

Accordingly, the += operator increases the value of a variable by the subsequent value, and the -= operator decreases it by the following value. The *=, /=, and %= operators therefore represent abbreviations of the following statements: z = z * [value], z = z / [value], and z = z % [value].

You can set the ++ and -- operators both before a variable (*prefix notation*) and after a variable (*postfix notation*). With the statement d = a++;, d receives the old value of a, and then a is increased. With the statement d = ++b;, b is increased first, and then d receives the increased value of b.

In this way, you can (unfortunately) also formulate very confusing expressions, the result of which will only be recognizable after long consideration, as you can see in the last example.

You can see the results of the assignments in Figure 2.7.

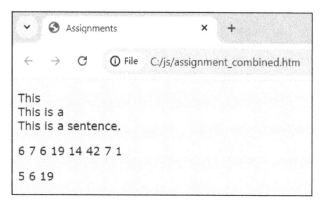

Figure 2.7 Assignment Operators

2.2.3   Entering Numbers

In this section, the prompt() method is used to enter two numbers, and the sum of these two numbers should then be output using alert().

The prompt() method returns a string. In Section 2.1.4, you have seen that the + operator handles strings and numbers differently, so it's necessary to first convert such an input into a number before it can be used to calculate a sum.

Here's a sample program:

```
...
<body>
    <script>
        const input = prompt("Number 1:");
        const z1 = parseFloat(input);
        const z2 = parseFloat(prompt("Number 2:"));
        const result = z1 + z2;
        document.write(z1 + " + " + z2 + " = " + result);
    </script>
</body></html>
```

Listing 2.7 numbers_input.htm File

The result of the first input is saved in the `input` variable. This is a string, and the `parseFloat()` function is used to convert a string into a floating point number (i.e., a number with decimal places). After this conversion, the `z1` variable contains a number.

The same happens with the second entry, only in a shortened form. The `prompt()` method returns a string for which the `parseFloat()` function is called directly, the two numbers are then added together, and the entire calculation is output. You can see an example in Figure 2.8 to Figure 2.10.

Decimal places are denoted by decimal points. If commas are used instead, the decimal places will be lost.

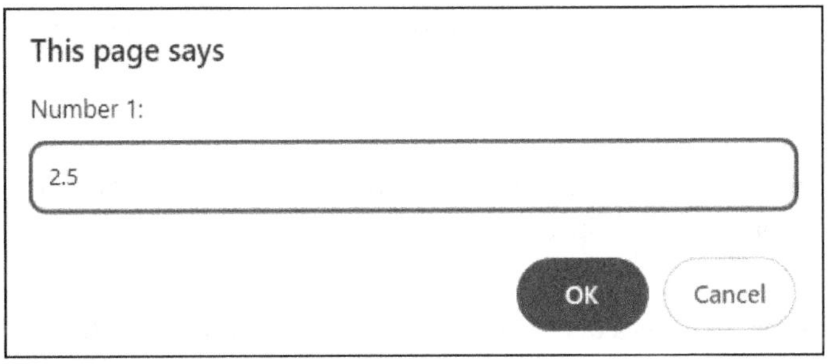

This page says
Number 1:
2.5
OK
Cancel

Figure 2.8 Entering First Number

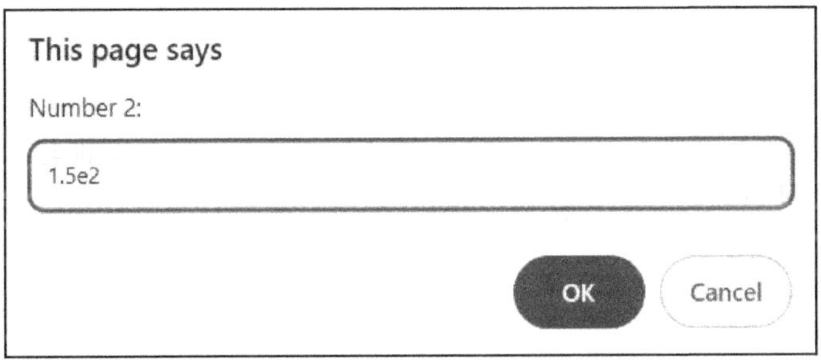

This page says
Number 2:
1.5e2
OK
Cancel

Figure 2.9 Entering Second Number

Numbers, input
File
C:/js/numbers_input.htm
2.5 + 150 = 152.5

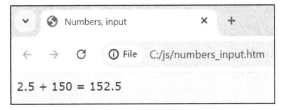

Figure 2.10 Output of Entire Calculation

> **Note**
>
> In the bonus chapter 2, which is included in the downloadable materials for the book at *www.rheinwerk-computing.com/5875*, you'll find an extension of the program in the *regexp_point.htm* file. It enables the input and output of numbers with a comma.

Numbers can also be entered in exponential format. If an input can't be converted into a number (e.g., "abc" or an empty input), parseFloat() returns the result NaN. In Section 2.3.5, we'll determine whether a number has been entered.

The parseInt() function works like the parseFloat() function in that it also cuts off the decimal places and returns an integer. Thus, the string "3.8" becomes the integer 3.

> **Note**
>
> The write() method is called for the document object, but the global parseFloat() and parseInt() functions are not called for a specific object.

Usually, parseInt() interprets a text as a decimal number (i.e., as a number with a base of 10). You could use a second parameter to specify a different base, so for example, the parseInt("101", 2) call results in the decimal value 5 because the text "101" is interpreted in base 2 (i.e., as a dual number).

> **Note**
>
> You'll find the u_number exercise in bonus chapter 1, section 1.2, found in the downloadable materials for this book at *www.rheinwerk-computing.com/5875*.

2.2.4   Number Systems

You'll usually use the decimal number system with the base of 10, but you can also use the toString() method to output numbers as a string that uses the digits of other number systems, such as the following:

- The dual number system (with base 2)
- The octal number system (with base 8)
- The hexadecimal number system (with base 16)

Here's an example:

```
...
<body><p>
  <script>
     const z = 27;
     document.write("Decimal: " + z + "<br>");
```

```
        document.write("Hexadecimal: " + z.toString(16) + "<br>");
        document.write("Octal: " + z.toString(8) + "<br>");
        document.write("Dual: " + z.toString(2) + "<br>");
    </script></p>
</body></html>
```

Listing 2.8 number_systems.htm File

The output of the program is shown in Figure 2.11.

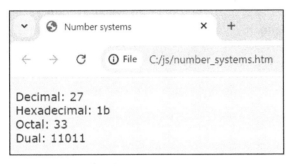

Figure 2.11 Number Systems

When the `toString()` method is called, the base of the desired number system gets specified.

In addition to the digits 0 to 9, the hexadecimal system uses the letters *a* to *f* (or *A* to *F*) as digits for the values from 10 to 15. Hexadecimal numbers start with the prefix 0x, octal numbers with 0o and dual numbers with 0b. The number 0x1b corresponds to the following value:

$$1 \times 16^1 + B \times 16^0 = 1 \times 16^1 + 11 \times 16^0 = 16 + 11 = 27$$

The octal system only uses the digits 0 to 7, so the number 0o33 corresponds to the following value:

$$3 \times 8^1 + 3 \times 8^0 = 24 + 3 = 27$$

The dual (or binary) system only uses the digits 0 and 1, so the number 0b11011 corresponds to the following value:

$$1 \times 2^4 + 1 \times 2^3 + 0 \times 2^2 + 1 \times 2^1 + 1 \times 2^0 = 16 + 8 + 2 + 1 = 27$$

2.3   Different Branches in a Program

Depending on certain conditions, different parts can be run through in a program. We also say that the program *branches out*. Branches are among the most important control structures, and in this section, you'll learn about different types of branches and their effects.

2.3.1   Branches with if

Branches can be created using an `if` statement. The `if` is followed by a condition in parentheses:

- If the condition is fulfilled, the subsequent statement or block of statements gets executed. You can recognize a block of statements by the curly brackets (e.g., {... }).
- If the condition is not fulfilled, the statement or the block of statements after the keyword `else` gets executed, if available.

Conditions are formed using comparison operators, and they result in truth values. The > operator stands for *greater than*, the < operator stands for *less than*, >= means *greater than or equal to*, and <= means *less than or equal to*. You can use == to check for equality and != to check for inequality.

You should pay particular attention to the double equals sign with the == operator. A common mistake is to confuse == with the = operator, which results in an assignment instead of a comparison. All comparison operators can be used for numbers, but only the last two—== and !=—can be used for strings.

You can use the ternary operator ?: to create a shortened branch, and a value is provided that can be assigned.

The following is an example with a total of six branches. These are numbered in the program and in the output for a better overview:

```
...
<body>
  <script>
    const a = 12, b = 7;

    // 1: Single branch
    if(a > b)
       document.write("<p>1: a is greater than b</p>");

    // 2: Single branch, with else
    if(a < b)
       document.write("<p>2: a is less than b</p>");
    else
       document.write("<p>2: a is not less than b</p>");

    // 3: Single branch, multiple statements in the block
    if(a > b)
    {
       document.write("<p>3: a is greater than b<br>");
       document.write("3: An additional line</p>");
    }
```

```
    // 4: Multiple branches
    if(a > b)
        document.write("<p>4: a is greater than b</p>");
    else if(a < b)
        document.write("<p>4: a is less than b</p>");
    else
        document.write("<p>4: a is equal to b</p>");

    // 5: Character strings
    const country = "Spain";
    if(country == "Spain")
        document.write("<p>5: Country is Spain</p>");
    if(country != "Spain")
        document.write("<p>5: Country is not Spain</p>");

    // 6: Ternary operator
    const greater = (a > b) ? a : b;
    document.write("<p>6: The greater number is " + greater + "</p>");
  </script>
</body></html>
```

Listing 2.9 if_else.htm File

In branch 1, the condition a > b is checked. This results in the truth value true for the numerical values specified here. The condition is therefore fulfilled, and for this reason, the document.write() method is called. If a is not greater than b, then the truth value is false and nothing is output here because there's no else.

Something is always output in branch 2, as there's also an else for the if.

A block of statements is executed in branch 3. If several statements are to be executed based on a condition, they must always be placed in curly brackets. If the brackets are missing, only the first statement will be executed, depending on the branch. The other statements are then executed in any case because it's no longer recognized that they should also belong to the branch.

In branch 4, all cases are checked, one after the other. If a is not greater than b, the system checks whether a is smaller than b. If this is also not the case, a and b are the same. With such multiple branches, you can also distinguish between three or more cases.

In branch 5, two strings are compared with each other.

In branch 6 of the program, the ternary operator is used to assign a value. If the condition in the brackets applies, the variable receives the value after the question mark (?); if not, then the variable receives the value after the colon (:).

The output of the program is shown in Figure 2.12.

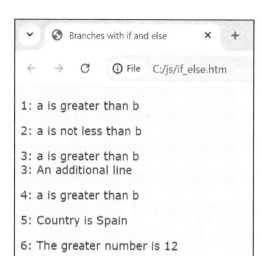

Figure 2.12 Six Branches

What will the output be if you set the value b = 17 in the program? Think through the answer on your own. Then change the program accordingly and check whether your assumption was correct. What happens if b = 12? Here too, first determine the result by thinking through and then use the modified program to check whether your assumption was correct.

Two Tips

- Write clear, easy-to-read programs. In the case of a branch, you should work with indentations after if and else, as in the previous program.
- Don't write a semicolon after the condition after an if or an else. Therefore, it should *not* look like this: if(a > b); or else;, which would result in the branch ending immediately and the subsequent statements always being executed. There could also be no output at all because the else is generated without its associated if. These typical rookie mistakes are hard to detect.

Note

You'll find the u_if exercise in bonus chapter 1, section 1.3, in the downloadable materials for this book at *www.rheinwerk-computing.com/5875*.

2.3.2 Requesting a Confirmation

The confirm() method of the window object provides a further interaction option in connection with a branch. After calling the method, a dialog box appears with a question and two buttons labeled **OK** and **Cancel**.

Here's an example:

```
...
<body>
   <script>
      const response = confirm("Do you want to carry out this action?");
      if(response)
         document.write("This action will be carried out");
      else
         document.write("This action will not be carried out");
   </script>
</body></html>
```

Listing 2.10 if_confirm.htm File

If you click the **OK** button, true will be returned. This is already sufficient for the branch, as a truth value is provided. You don't need a comparison operator, and the if(response == true) query would lead to the same result.

If you click the **Cancel** button, false will be returned and the else branch will be run through.

In Figure 2.13, you can see the dialog box with the question.

Figure 2.13 Confirmation with "confirm()"

2.3.3   Linking Multiple Conditions

If multiple conditions are to be linked together, you can use the logical operators && for the logical AND or || for the logical OR. With the logical AND, all individual conditions must be true for the entire condition to be true. With the logical OR, one true condition is sufficient for the entire condition to be true.

There's also the logical operator ! for the logical NOT, which reverses the truth value of a condition. True becomes false, and false becomes true.

Here's an example:

```
...
<body><p>
   <script>
```

```
    const a = 12;

    if(a >= 10 && a <= 20)
        document.write("1: between 10 and 20</br>");
    else
        document.write("1: not between 10 and 20<br>");

    if(a < 10 || a > 20)
        document.write("2: not between 10 and 20<br>");
    else
        document.write("2: between 10 and 20<br>");

    if(!(a >= 10 && a <= 20))
        document.write("3: not between 10 and 20");
    else
        document.write("3: between 10 and 20");
  </script></p>
</body></html>
```

Listing 2.11 if_linked.htm File

In the first case, the system checks whether the number is greater than or equal to 10 *and* less than or equal to 20 (i.e., whether it's within the number range from 10 to 20).

The second case tests whether the number is less than 10 *or* greater than 20 (i.e., outside the number range).

In the third case, the ! operator is used to determine whether the number is *not* in the number range between 10 and 20.

In Figure 2.14, you can see the outputs of the various links if the numerical value 12 is checked in each case.

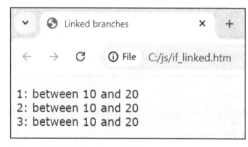

Figure 2.14 Linked Conditions

What will the output be if you set the value a = 25 in the program? Think through the answer on your own. Then change the program accordingly and check whether your assumption was correct. What happens when a = –15? Here too, first determine the result by thinking through and then use the modified program to check whether your assumption was correct.

Two Tips

- The a >= 10 || a <= 20 condition is fulfilled by every number, while the a < 10 && a > 20 condition is not fulfilled by any number. You should avoid these links.

- The so-called short-circuit *behavior* is defined for the logical operators in JavaScript. This is important for the use of functions, and in Section 2.6.7, you can see an example of this.

Note

You can find the u_linked exercise in bonus chapter 1, section 1.4, in the downloadable materials for this book at *www.rheinwerk-computing.com/5875*.

2.3.4   Linking and Assigning

Since ECMAScript 2021, there have been the two operators, &&= and ||=, which combine a logical link with an assignment. This rarely used option is illustrated in the following program, where three truth values are used:

```
...
<body><p>
   <script>
      let a = true, b = true, c = false;
      document.write("a: " + a + ", b: " + b + ", c: " + c + "<br>");
      a &&= b;        document.write("a after a&&=b : " + a + "<br>");
      a &&= c;        document.write("a after a&&=c : " + a + "<br>");
      a ||= c;        document.write("a after a||=c : " + a + "<br>");
      a ||= b;        document.write("a after a||=b : " + a);
   </script></p>
</body></html>
```

Listing 2.12 assignment_linked.htm File

Each of the results of the individual statements is each assigned to variable a. The output of the program is shown in Figure 2.15.

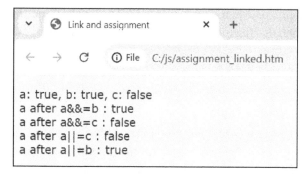

Figure 2.15 Linking and Assignment

2.3.5   Checking the Input of Numbers

In this section, an entered number gets checked using multiple branches. During the course of the program, you'll become familiar with the following useful functions and methods:

- The isNaN() function determines whether a variable *doesn't* contain a valid number.
- The Math.random() method generates a random number that's greater than or equal to 0 and less than 1.
- The Math.abs() method calculates the absolute value of a number (i.e., the number without its sign). The value –5 becomes +5, and the value 5 remains +5.
- The random() and abs() methods are methods of the Math object, which you'll get to know in more detail in Chapter 6, Section 6.3.

Here's the sample program:

```
...
<body><p>
   <script>
       const number_random = Math.random();
       const input = prompt("Please enter a number from 0 to 1, without 1");
       const number_input = parseFloat(input);
       const distance = Math.abs(number_random - number_input);

       if(input == null)
          document.write("Cancel");
       else if(input == "")
          document.write("No input");
       else if(isNaN(input))
          document.write("No valid number");
       else if(number_input < 0 || number_input >= 1)
          document.write("No number in the valid range");
```

```
      else if(distance < 0.1)
         document.write("You are close");
      else
         document.write("You are far off");

      document.write("<br><br>Random: " + number_random + "<br>");
      document.write("Input: " + input + "<br>");
      document.write("isNaN(): " + isNaN(input) + "<br>");
      document.write("Number: " + number_input + "<br>");
      document.write("Difference: " + (number_random - number_input) + "<br>");
      document.write("Distance: " + distance);
   </script></p>
</body></html>
```

Listing 2.13 if_else_input.htm File

In the first part of the program, a random number gets generated and saved. Due to the use of prompt(), you are prompted to make an entry, and this input is saved both as a string and as a number.

The second part of the program contains multiple branches. If the first condition doesn't apply, the second will be checked. If this doesn't apply either, the third will be checked, and so on.

The **Cancel** button causes the value null to be saved (see also Section 2.1.3). If nothing is entered and then the **OK** button gets clicked, an empty string will be saved. You can query both using the == operator.

The isNaN() function stands for *is not a number.* If the input doesn't contain a valid number, the method returns true; if it does, then the method returns false. The subsequent linked condition checks whether the number is within the valid range.

Then the difference between the random number and the entered number gets determined. The absolute value is calculated from this difference, and if the absolute value is less than 0.1, the input is close to the actual number.

Why does the absolute value actually have to be calculated? If we assume that the random number is 0.34 and the input is 0.3, then the difference is 0.04 and you don't need to calculate the absolute value. However, if the random number is 0.34 and the input is 0.5, then the difference is −0.16, thus less than 0.1. This would be "close" without calculating the absolute value, but that is not the case. The Math.abs() method calculates +0.16, and that is no longer "close". So the absolute value should be calculated.

Finally, the program provides an output of all important values. Figure 2.16 and Figure 2.17 show possible outputs.

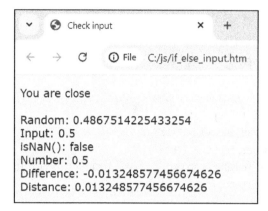

Figure 2.16 Output after Entering 0.5 with Decimal Point

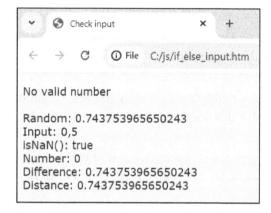

Figure 2.17 Output after Entering 0,5 with Comma

> **Note**
> The Math.random() method is suitable for many purposes in simulations or games, but not for encryption. This would require a better random number generator, such as that offered by the crypto object (see Chapter 6, Section 6.3.3).

2.3.6   Checking the Value and Type

The === and !== operators perform checks that are a little more precise than those from the == and != operators. For a condition with ===, both the value of the variable and the type of this value must match. The system therefore also tests whether the value is a number, a string, or a truth value, for example.

The typeof operator is also interesting in this context because it provides you with the type of a value or a variable.

Here's a program that uses these operators:

```
...
<body><p>
    <script>
        const a = 4711;
        const b = "4711";
        const c = 4711;
        const d = true;

        document.write("<p>a: " + a + ", " + typeof a + "<br>");
        document.write("b: " + b + ", " + typeof b + "<br>");
        document.write("c: " + c + ", " + typeof c + "<br>");
        document.write("d: " + d + ", " + typeof d + "<br><br>");

        if(a == b)
            document.write("a == b<br>");
        if(a === c)
            document.write("a === c<br>");

        if(a === b)
            document.write("a === b");
        else
            document.write("Not a === b");
    </script></p>
</body></html>
```

Listing 2.14 if_exact.htm File

In Figure 2.18, you can see the name, the value, and (thanks to typeof) the type of the value of the four variables that are used in the program. In the case of a number, typeof returns the designation number, in the case of a string, the designation string and in the case of a truth value, Boolean.

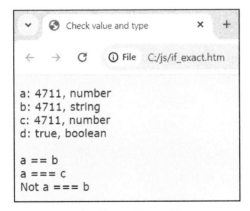

Figure 2.18 Checking Value and Type

The a, b, and c variables all contain the value 4711, but in b as a string. The true value is always returned for all comparisons between these variables using the == operator. If the === operator is used, this is only successful when comparing a and c.

2.3.7   Priority of Operators

The operators have different priorities in JavaScript. If multiple operators are used in a statement, the parts of the statement with the higher priority will be executed first. If their priority is the same, the statement is executed from left to right.

Table 2.1 shows the operators used so far in order of their priority.

Priority	Operators	Description
1 (highest)	( )	Parentheses
2	! , −, ++, --	Logical not, negative sign, increment, decrement
3	**	Exponentiation
4	*, /, %	Multiplication, division, modulo
5	+, -	Addition, subtraction
6	<, >, <=, >=	Less than, greater than, less than/equal to, greater than/equal to
7	==, !=, ===, !==	Equal to, not equal to, equal to with type comparison, not equal to with type comparison
8	&&	Logical AND
9	||	Logical OR
10 (lowest)	=, +=, -=, *=, /=, %=	Assignments

Table 2.1 Priority of Operators

2.3.8   Branches with switch

The switch keyword provides another option for creating a branch. In some cases, a branch with switch is clearer and easier to read than a branch with if.

Here's an example:

```
...
<body>
  <script>
    const country = prompt("Please enter a country:");
    let city;
```

```
switch(country)
{
  case "Italy":
      city = "Rome";
      break;
  case "England":
  case "Wales":
  case "Scotland":
      city = "London";
      break;
  default:
      city = "Not known";
}

document.write("Capital of the country: " + city);
  </script>
</body></html>
```

Listing 2.15 switch.htm File

The name of a country is supposed to be entered, and the entered text gets saved and then analyzed. The `switch` keyword is followed by the saved text in parentheses.

This is followed by a block in curly brackets. Within the block, the cases are differentiated using the `case` keyword, and a check is run from top to bottom to determine whether the value matches one of these cases. If it does, all subsequent statements are executed up to the next `break` statement.

If `Italy` was entered, `Rome` is provided as the capital and the `switch` block is exited.

If `England` was entered, `London` is returned as the capital because the next `break` statement follows after that. The same applies to `Wales` and `Scotland`. By using `switch`, you therefore have the option of combining multiple cases.

If none of the cases apply, you can use the `default` keyword to catch all remaining cases. As with an `if` without `else`, however, this case doesn't necessarily have to exist.

You can use a `switch` branch to examine a specific number or a specific string, but it's not possible to specify ranges such as `case < 10` or `case >= 10 && <= 20`.

2.4 Repeating Program Sections

In addition to branches, *loops* are important control structures. Many processes that are repeated in an identical or a similar way can be programmed efficiently with loops, using the ability of a computer to perform a large number of steps in a short time.

2.4.1 Loops with for

You can use a loop with for in the following cases:

- You know how often a program section should be repeated.
- Part of the program is to be repeated for a regular sequence of numbers that run from a start value to a final value.

Usually, a variable called a *loop variable* is used to control for loops. The structure of a for loop is divided into three parts:

- A statement for the start value of the loop variable
- A condition that must apply to the loop variable for the entire duration of the loop
- A statement for changing the loop variable after a single pass of the loop

The following example shows five different loops:

```
...
<body>
   <script>
      // 1: Upward
      document.write("<p>1: ");
      for(let i=1; i<=5; i++)
         document.write(i + " ");

      // 2: Downward
      document.write("<br>2: ");
      for(let i=20; i>=10; i--)
         document.write(i + " ");

      // 3: With decimal number
      document.write("<br>3: ");
      for(let i=3; i<=4; i+=0.2)
         document.write(i + " ");

      // 4: With decimal number, optimized
      document.write("<br>4: ");
      for(let i=3; i<=4.1; i+=0.2)
         document.write(i.toFixed(1) + " ");

      // 5: With break and continue
      document.write("<br>5: ");
      for(let i=10; i<=50; i++)
      {
         if(i>=16 && i<=24)
            continue;
```

```
        if(i>30)
            break;
        document.write(i + " ");
    }
    document.write("</p>");
</script>
</body></html>
```

Listing 2.16 for.htm File

In loop 1, the loop variable i is given the start value 1. The following applies as a condition for the entire loop: It runs as long as i is less than or equal to 5, and after each run, the value of i is increased by 1 using the assignment operator, ++. This results in a sequence of numbers from 1 to 5, in increments of 1. The start value, condition, and change are separated from each other by a semicolon.

There's only one statement within loop 1: the output of the loop variable. In Figure 2.19, you can see the regular sequence of numbers. If multiple statements are supposed to be executed within a loop, they must be in a block with curly brackets, as in a branch.

In contrast to loop 1, loop 2 runs downward. It starts at 20 and runs as long as the value of i is greater than or equal to 10. After each run, the value of i is reduced by 1, and this results in the following number sequence: 20, 19, 18, ... 10.

The various parts of a loop must be coordinated with each other. A loop with for(let i = 10; i >= 5; i++) runs endlessly, while a loop with for(let i = 10; i <= 5; i--) never runs at all. You should avoid such loops at all costs.

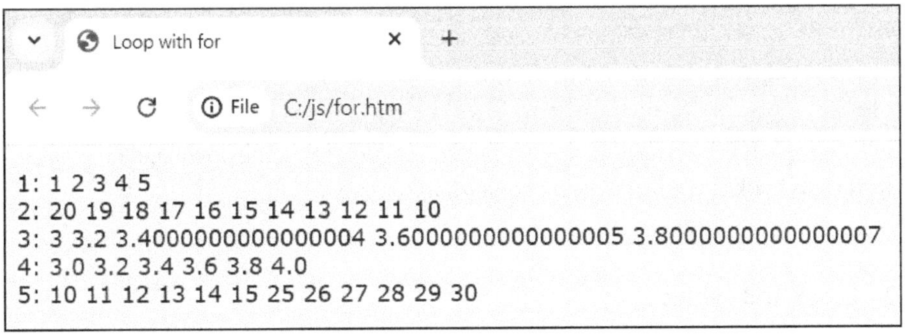

```
Loop with for                        ×    +

←   →   C    ⓘ File   C:/js/for.htm

1: 1 2 3 4 5
2: 20 19 18 17 16 15 14 13 12 11 10
3: 3 3.2 3.4000000000000004 3.6000000000000005 3.8000000000000007
4: 3.0 3.2 3.4 3.6 3.8 4.0
5: 10 11 12 13 14 15 25 26 27 28 29 30
```

Figure 2.19 Five Different "for" Loops

In loop 3, you can see that a loop variable can also be a number with decimal places. In contrast to integers, numbers with decimal places can't be saved with mathematical precision, and that gives rise to two problems:

- An output results in unsightly small deviations, as you can see in Figure 2.19 from the value 3.4 onwards.

■ The loop doesn't run to the desired end of 4. After the increase of 0.2, the value of i is just above 4, and a value above 4 is no longer reached due to the condition i<=4.

Both problems can be solved, as you can see in loop 4 in Figure 2.19. First, a value is entered within the condition that's clearly above the last desired value. This can be a value that's half an increment higher, for example. In addition, the toFixed() method is used to output the numbers. This is a method of the Number object that you'll get to know in Chapter 6, Section 6.3. The toFixed() method rounds a value to the desired number of decimal places (e.g., for the output).

Loop 5 contains multiple statements that must be placed in a block with curly brackets. Loop 5 contains the continue and break statements.

The continue statement causes the current loop run to be ended immediately and the next run to continue. This means that the values from 16 to 24 are not output, as you can see in Figure 2.19.

The break statement leads to an immediate termination of the entire loop, which would otherwise run to 50. Both statements are normally used within a branch that's intended to handle certain special cases.

The scope of the loop variable i is limited to the respective loop itself, thanks to the declaration with let.

Note

If variables are declared with let or const within a block of statements, they are only known within the block. This makes them easier to control and harder to change accidentally.

Note

You'll find the u_for exercise in bonus chapter 1, section 1.5, in the downloadable materials for this book at *www.rheinwerk-computing.com/5875.*

2.4.2   Loops and Tables

Loops can be used to easily create dynamic tables. Here are some examples:

```
...
<body>
   <script>
      document.write("<table>");
      for(let z=0; z<3; z++)
         document.write("<tr><td>Z" + z + "</td></tr>");
      document.write("</table>");
```

```
    document.write("<br>");

    document.write("<table><tr>");
    for(let s=0; s<3; s++)
        document.write("<td>S" + s + "</td>");
    document.write("</tr></table>");
    document.write("<br>");

    document.write("<table>");
    for(let z=0; z<3; z++)
    {
        document.write("<tr>");
        for(let s=0; s<5; s++)
            document.write("<td>Z" + z + "/S" + s + "</td>");
        document.write("</tr>");
    }
    document.write("</table>");
  </script>
</body></html>
```

Listing 2.17 for_table.htm File

The output of the program is shown in Figure 2.20.

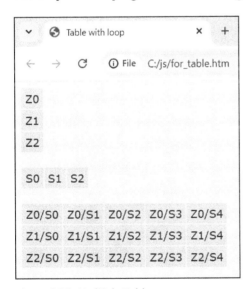

Figure 2.20 Multiple Tables

The first step is to create a table with three rows. Each row contains a table cell in which the current number of the row is displayed, and the numbering starts at 0. You'll often encounter this type of numbering in connection with fields (Section 2.4.3).

Then, a table with one row gets created. This row contains three table cells, each of which displays the current number of the table cell.

Finally, there's a table with three rows and five columns. The number of the current row and the current column is displayed in each table cell.

Two nested loops are used and are controlled with different variables:

- Initially, the z variable of the outer loop has the value 0. The inner loop is then run through a total of five times using the s variable and the values 0 to 4 to create the five cells of a row.

- Then, z receives the value 1, and five cells are created again, and so on.

> **Note**
>
> You'll find the u_table exercise in bonus chapter 1, section 1.6, in the downloadable materials for this book at *www.rheinwerk-computing.com/5875*.

2.4.3   Loops and Fields

Fields are used to record large amounts of thematically related data. They consist of individual elements, and all elements of a field can be reached quickly using loops. We have already introduced the first simple fields at this point, and you can find out more about fields in Chapter 6, Section 6.1.

Here's a sample program:

```
...
<body>
   <script>
      let person;
      person = [ "Peter", "Monica", "John" ];

      document.write("<table>");
      for(let z=0; z<3; z++)
         document.write("<tr><td>" + person[z] + "</td></tr>");
      document.write("</table>");
      document.write("<br>");

      const age = [ ["Peter", 37], ["Monica", 35],  ["John", 32] ];

      document.write("<table>");
      for(let z=0; z<3; z++)
      {
         document.write("<tr>");
         for(let s=0; s<2; s++)
```

```
                document.write("<td>" + age[z][s] + "</td>");
            document.write("</tr>");
        }
        document.write("</table>");
    </script>
</body></html>
```

Listing 2.18 for_array.htm File

An *array* can be created using rectangular brackets. Within the brackets, there are the individual elements of the field, separated by commas. Usually, the elements of a field belong together thematically and are of the same type (e.g., numbers, strings, truth values).

The person variable represents a reference to the field after the field has been assigned, and the field and its elements can be accessed using the reference. The name of the field is person, and you can change the individual elements of the field as well as assign a different field to the variable.

The individual elements of the field can be identified by their sequential number: the so-called *index*, which is specified in rectangular brackets after the name of the field. The first element can be accessed via person[0], the second element via person[1], and so on. Indexing always starts at 0.

This relationship is used within the for loop to output all elements of the field in a table. The loop variable z runs through all indexes of the field, and for a field with three elements, these are the indexes 0, 1, and 2.

You can see the output of the table in Figure 2.21.

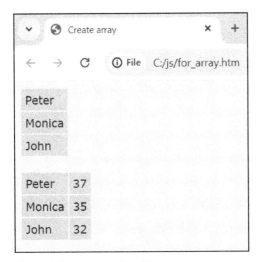

Figure 2.21 Loops and Fields

Fields can in turn contain fields as elements, which allows you to create multidimensional fields.

In this example, each of three sets of data on three people is in a one-dimensional field. The three one-dimensional fields are again written within rectangular brackets and separated by commas. In this way, a two-dimensional field is generated, and it can be clearly displayed in a two-dimensional table.

Regardless of the number of dimensions, you can also use an unchangeable variable as a reference to a field. However, only the reference is unchangeable. You can still change the individual elements of the field, but you can't assign a different field to the variable.

The individual elements of a two-dimensional field can only be accessed via two indexes, each of which is contained within rectangular brackets. You can access the name of the first person via age[0][0], the person's age via age[0][1], the name of the second person via age[1][0], and so on.

This relationship is used within the nested for loop to output all elements of the field in a table. The loop variable z runs through the indexes of the rows of the field, while the loop variable s runs through the indexes of the columns of the field.

You can also see the output of the two-dimensional table in Figure 2.21.

2.4.4 Loops with while

A loop with while depends on a condition that's checked at the start of the loop.

In the following example, random numbers are added together, and the while loop runs as long as the sum of the numbers hasn't reached the value 4. During the development of the program, the values are not known due to their random nature, so it's not known how many numbers have to be added (i.e., how often the loop will run).

Here's the program:

```
...
<body>
   <p><script>
      let sum = 0;
      while(sum < 4)
      {
         summe += Math.random();
         document.write(sum + "<br>");
      }
   </script></p>
</body></html>
```

Listing 2.19 while.htm File

The sum of the numbers is initially set to 0 for two reasons:

- At the beginning of the loop, the sum is compared with the value 4, and a comparison only makes sense if all the values involved are known.

- Within the loop, the sum is given a new value, which is added to the old value. For this reason, an old value of 0 is necessary at the beginning.

As long as the sum is less than 4, the loop will be run through. Since the loop includes multiple statements, you must use curly brackets. Within the loop, the sum is increased by a random value and output, and you can see a possible output in Figure 2.22.

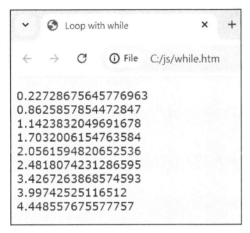

Figure 2.22 "while" Loop for Adding Random Numbers

2.4.5   Loops with do … while

A loop with do … while depends on a condition that's checked at the end of the loop. This loop is run through at least once, unlike the while loop.

The following program contains a mental arithmetic task. Two random numbers between 1 and 20 are to be added together, and the program prompts you until the correct result has been entered.

As with the while loop, it's not known during the development of the program how often the loop must be run through. However, at least one attempt is required, which is why the do … while loop is used. A do … while loop is also referred to as a *read-in loop*, as it's often used for the controlled input of a value:

- The input is entered at least once.
- If the entry doesn't meet the specifications, it will be repeated.

Here's the program:

```
...
<body>
   <p><script>
      const number_random = Math.random() * 20 + 1;
      const a = Math.floor(number_random);
      const b = Math.floor(Math.random() * 20 + 1);
      const sum = a + b;
      let number_input;

      do
      {
         const input = prompt(a + " + " + b + " = ");
         number_input = parseInt(input);
         if(number_input != sum)
            alert("Please try again");
      }
      while(number_input != sum);

      document.write(a + " + " + b + " = " + sum);
   </script></p>
</body></html>
```

Listing 2.20 do_while.htm File

The random number generator returns a number between 0 and 1, excluding the 1. If this number is multiplied by 20, the result is a number from 0 to 20, excluding the 20. If we add 1, the result is a number between 1 and 21, excluding the 21, as is the case here for the number_random variable.

The floor() method of the Math object cuts off all decimal places of a number, so the number 3.8 would become the number 3. You can find out more about the Math object in Chapter 6, Section 6.3. The Math.floor() method is first called with the number_random variable, and it's then called directly, without intermediate storage, for the random number between 1 and 20. In this way, the process can be shortened.

The text entered is saved in the number_input variable, and it's then converted into an integer using parseInt(). This is followed by a condition that's used twice: is the integer different from the correct result? If this is the case, a message gets displayed, and the do... while loop continues to run.

If the correct result has been entered, the condition no longer applies, the loop ends, and the task and the correct result are displayed again in the document.

In Figure 2.23, you can see the task and a possible incorrect entry. The program's response to this is shown in Figure 2.24.

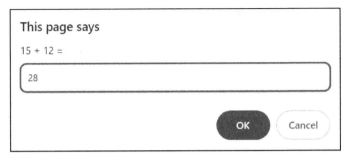

Figure 2.23 Task and Incorrect Input

Figure 2.24 Response of Program

Three Tips

- At the start of the programming work, the condition of the while or do … while loop and the condition in the second part of the for loop present a certain challenge. Remember that this is a run condition and not a termination condition, so it's not the case that the loop terminates as soon as the condition is met. The loop runs *as long as* the condition applies.

- The beginning of a for loop or a while loop must not be followed by a semicolon, so it should *not* look like for(let i = 1; i <= 5; i++); or while(a > b);. This would result in the loop terminating immediately and the subsequent statements only being executed once. This error is difficult to detect.

- In contrast, the end of a do … while loop must be followed by a semicolon.

2.4.6   A Game for Memory Training

In this section, we develop a game that people can use to train their memory. As no new language elements are added, you can skip this section, if you want. However, it's interesting to see how the language elements you have learned about so far interact with each other in the game.

Here are the rules of the game, which also appear after calling the file in the browser: Memorize the number sequences that are about to appear and then enter them. First, three sets of three digits appear, then three sets of four digits, and so on. As soon as you have entered a sequence of digits incorrectly, you have reached the end of the training and will be scored. When you refresh the website, the training starts again from the beginning.

Here's the program:

```
...
<body>
   <p><script>
      let counter = 0, length = 3, text, input, output = "";

      alert("Memory training:\n\nMemorize the number sequences "
         + "that appear and then enter them. First three sets of "
         + "three digits appear, then three sets of four digits "
         + "and so on. As soon as you have entered a sequence of "
         + "digits incorrectly, you have reached the end of the training "
         + "and will be scored. When you refresh the website, "
         + "the training starts again from the beginning.");

      do
      {
         counter++;
         if(counter>3)
         {
            length++;
            counter = 1
         }

         text = "";
         for(let i=1; i<=length; i++)
            text += Math.floor(Math.random() * 10);
         alert("Sequence of digits: " + text);

         input = prompt("Your input");
         output += text + "<br>";
      }
      while(input == text);

      let quantity = length - 1;
      if(quantity < 3)
         quantity = 0;
      output += "You were able to memorize " + quantity + " digits";
      document.write(output);
```

```
    </script></p>
</body></html>
```

Listing 2.21 memory.htm File

The random number sequences are stored in the text variable. The length variable contains the current length of the digit sequence, and counter contains information on whether it's the first-, second-, or third-digit sequence of the same length.

The do ... while loop runs as long as the input and the sequence of digits match. The value of counter gets increased each time, and if the value exceeds 3, it will be reset to 1. At the same time, the value of length increases, which ensures that a sequence of digits of the same length appears three times.

The for loop is used to compile a random sequence of digits of the current length. This is issued and must be remembered, and it must then be entered correctly again.

After the end of the loop (i.e., after an incorrect entry), the result appears. It's a list of all digit sequences that have occurred up to that point, and at the end, the number of digits successfully memorized and re-entered three times will get displayed. In Figure 2.25, you can see a possible result that you'll hopefully surpass.

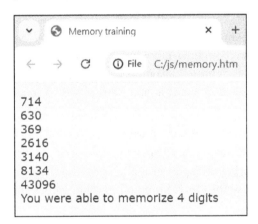

Figure 2.25 Memory Training

2.5   Finding and Avoiding Errors

During the development of a program, you'll make the odd mistake. This is normal and part of everyday life when you write programs. It's more important to know how to find and rectify errors quickly—and it's even better to know how to avoid mistakes.

These may be *syntax errors* (i.e., notations of incorrect or incomplete JavaScript statements), but there may also be *logical errors* (in which your program is running perfectly but is not producing the results you expect).

2.5.1   Developing a Program

You should proceed step by step. First, give some thought to how the entire program should be structured, and do it on paper. Which parts should it consist of, in sequence? Don't try to write the entire program with all its complex components at once. This is the biggest mistake you can make at the beginning—or sometimes even later!

First, write a simple version of the first part of the program, and then test it. Don't add the subsequent program section until you've performed a successful test, and test again after each change so that if an error occurs, you'll know that it was caused by the last change. After the last addition, you will have created a simple version of your entire program.

Now, change part of your program to a more complex version. In this way, you'll make your program more complex step by step until you have finally created the entire program as it corresponds to your initial considerations on paper.

Sometimes, one or two changes to your design arise during the actual programming process. This is not a problem as long as the entire structure doesn't change, but if it does change, then you should return to your papers briefly and reconsider the structure. This doesn't mean that you have to delete the previous program lines, but possibly, you can just change them a little and arrange them differently.

Write your programs clearly. If you are thinking about how you can write three or four specific steps of your program at once, turn them into individual statements that are executed one after the other. This simplifies any troubleshooting process, and if you (or another person) change or expand your program at a later date, it will be much quicker to start building the program.

I have already mentioned some typical errors in connection with branches and loops. Have you closed all the brackets and parentheses you have opened? That's a classic programming error that often leads to nothing being displayed at all, especially with Java-Script. You have also already seen how to use the alert() method to check values and search for logical errors.

You can put individual parts of your program in comment brackets to determine which part of the program runs without errors and which part does contain errors.

2.5.2   Finding Errors Using onerror

One way to find errors is to use the onerror event. As soon as certain errors occur, the event gets triggered, and a function is called that provides you with information about the error. Let's take a look at the following example:

```
...
<body><p>
    <script>
```

```
      const y = 42;
      document.write(x + "<br>");
      document.write(y + "<br>");
   </script></p>
</body></html>
```

Listing 2.22 onerror.htm File, First Version

The values of the two variables should be displayed on the screen, but the x variable never gets declared. The attempt to output x results in the program being aborted, which means that the value of the correctly declared y variable will no longer be output either. The screen remains blank and the cause remains unknown.

So, let's now extend the program for troubleshooting purposes:

```
... <head> ...
   <script>
      function error_handling(error, file, line)
      {
         alert("Error: " + error + "\nFile: " + file
            + "\nLine: " + line);
      }
   </script>
</head>
<body><p>
   <script>
      onerror = error_handling;
      const y = 42;
      document.write(x + "<br>");
      document.write(y + "<br>");
   </script></p>
</body></html>
```

Listing 2.23 onerror.htm File, Second Version

Section 2.6 explains how you can modularize programs using your own functions, but let me provide just a brief explanation at this point.

The onerror event occurs when an error occurs, and as a result, the function that was assigned to the event gets called. In this example, that's the error_handling() function. Functions are usually defined in the head of the document, and the error_handling() function is provided with information about the error that has occurred, which you can output using alert().

The message for our example is shown in Figure 2.26, and it's referred to as an **Uncaught ReferenceError**. A *reference error* occurs if no reference (i.e., no declared name) exists for a variable. Errors can be *caught*, but this error is not caught. It occurs in line 18 of the *onerror.htm* file, and thanks to this message, the cause of the error can be found and the problem can be solved.

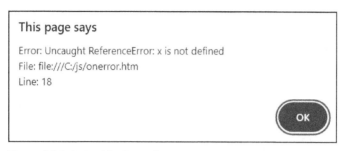

Figure 2.26 Output in Case of Error

2.5.3   Exception Handling Using try ... catch

Exception handling using try ... catch enables you to prevent your program from aborting. If you suspect that errors may occur in certain parts of the program due to inputs or other causes, you can place the relevant program parts in a try block.

If an error actually occurs, the program branches to a catch block in which the error is caught. An error message can be displayed there, or alternative statements can be run that avoid the consequences of the error.

If no error occurs, the statements in the try block are processed normally and the catch block is skipped. There may also be a finally block that's processed in each case, and this applies whether an error has occurred or not. Curly brackets must always be written, even in the case of a single statement.

Here's an example:

```
...
<body><p>
   <script>
      const y = 42;

      try
      {
         document.write(x + "<br>");
         document.write(y + "<br>");
      }
      catch(e)
      {
         alert(e);
      }
```

```
      finally
      {
         document.write("This will definitely be done<br>");
      }
   </script></p>
</body></html>
```

Listing 2.24 try_catch.htm File

The values of the two variables are to be output in the `try` block. The missing declaration of x is noticed, the program branches to the `catch` block, and information on the error that has occurred is transmitted in an error object. It has become common practice to designate such an error object with e (for *error*), and you can output the object using `alert()`.

The output in Figure 2.27 contains a **ReferenceError**. In contrast to the previous program, the error has been intercepted, which is why **Uncaught** doesn't appear and the program continues to run.

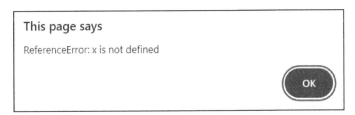

Figure 2.27 Catching Errors Using try … catch

2.5.4   Throwing Exceptions Using throw

You can also create your own error situations in which an exception can be thrown by using `throw`. You benefit from the exception handling process, which ensures that the program branches to the `catch` block.

In the following example, a number greater than 0 is to be entered. There are supposed to be two situations that count as errors:

- The entry is not a valid number.
- A number is entered that's too small.

In both cases, a matching error message is to be displayed.

Here's the program:

```
...
<body><p>
   <script>
```

```
    const number = parseFloat(prompt("Please enter a number > 0"));

    try
    {
        if(isNaN(number))
            throw "No valid number";
        if number <= 0:
            throw "Number too small";
        document.write("Number: " + number);
    }
    catch(errorObject)
    {
        alert(errorObject);
    }
  </script></p>
</body></html>
```

Listing 2.25 throw.htm File

If one of the two situations occurs, then it is recognized thanks to the branch. The throw statement ensures that the program branches directly to the catch block, and at the same time, an associated error message is sent to the catch block. This message is output, and in Figure 2.28, you can see the message that displays if you enter the value -5.

Figure 2.28 Custom Exception Thrown

2.5.5   Debugging a Program

In modern browsers, you have the option of debugging your programs. Debugging helps you to find logical errors, and you can run your programs step by step and view the values of all variables at any point in time. We show this in this section using the example of the Google Chrome browser.

As an example, a program in the *chrome_debug.htm* file is used in which two numbers are added. In Figure 2.29, you can see the default output of the program.

When you initiate debugging, it's assumed that the browser is open with the named program. Open the browser menu using the three dots in the top right-hand corner and select the **More tools • Developer tools** menu item. Then, the developer tools are displayed, and they may offer many different options.

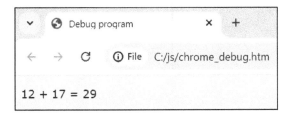

Figure 2.29 Program, Normal Procedure

For debugging, you want to select the **Sources** tab at the top of the screen. The program code is displayed as shown in Figure 2.30, and the individual lines are preceded by line numbers.

```
1    <!DOCTYPE html><html lang="en-us">
2    <head>
3        <meta charset="UTF-8">
4        <title>Debug program</title>
5        <link rel="stylesheet" href="js5.css">
6    </head>
7    <body>
8    <p><script>
9        const a = 12;
10       const b = 17;
11       const c = a + b;
12       const output = a + " + " + b + " = " + c;
13       document.write(output);
14   </script></p>
15   </body>
16   </html>
17
```

Figure 2.30 Program Code with Line Numbers

After you click on one of the line numbers, a *breakpoint* is created for this line. A second click removes the breakpoint. In Figure 2.31, you can see that breakpoints have been created next to the lines in which the b and output variables receive their values.

After reloading the document, the program doesn't run through completely but stops before executing the line in which the first breakpoint was set. The a variable has already been assigned a value at this point, and the other variables haven't yet.

```
 1    <!DOCTYPE html><html lang="en-us">
 2    <head>
 3      <meta charset="UTF-8">
 4      <title>Debug program</title>
 5      <link rel="stylesheet" href="js5.css">
 6    </head>
 7    <body>
 8    <p><script>
 9      const a = 12;
10      const b = 17;
11      const c = a + b;
12      const output = a + " + " + b + " = " + c;
13      document.write(output);
14    </script></p>
15    </body>
16    </html>
17
```

Figure 2.31 Two Breakpoints Set

You can monitor the values of the variables farther down in the **Watch** tab. To show the variables, click the button with the **+** sign, enter the name of the respective variable, and press Enter.

In Figure 2.32, you can see the status after stopping the program, after all four variables of the program have been displayed.

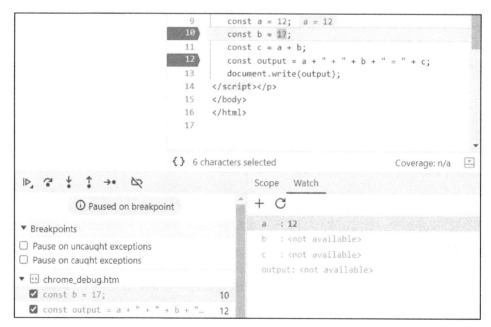

Figure 2.32 Program Is before Line 10

You can continue the execution of the program via the **Resume Script Execution** button, which you can see on the left above the **Paused on breakpoint** text in Figure 2.32. (Alternatively, you can press F8.) You use this button to continue the program to the next breakpoint, if available, or to the end of the program. In Figure 2.33, you can see the situation when you reach the second breakpoint.

Up to this point, the b and c variables have also been assigned values, and after the next resume, the program runs to the end and the output appears. After removing the breakpoints, the program runs again without stopping.

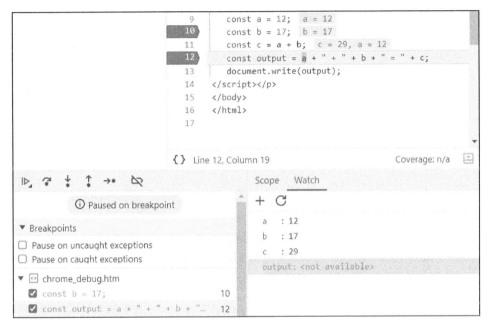

Figure 2.33 Program Is before Line 12

2.6   Custom Functions

You can break down programs into smaller, more manageable parts by using your own functions. This process is referred to as *modularization*, and a modularized program is easier to develop and change than a single large program. You can also swap out certain processes that you need again and again into functions. You only need to define them once, and then you can call them (i.e., use them) as often as you like.

Methods are functions that can only be run in connection with certain objects; more on this will follow in Chapter 3. You have already enjoyed the benefits of modularization because you have already used some predefined functions and methods. These were, for example, the object-independent parseInt(), parseFloat(), and isNan() functions, as well as the document.write() and Math.random() methods.

2.6.1 Simple Functions

Let's start with the simple functions. In this section, you'll see a function that produces exactly the same result every time it's called.

Here's the sample program:

```
...  <head> ...
   <script>
      function hello()
      {
          document.write("<p>Hello world</p>");
      }
   </script>
</head>
<body>
   <script>
      hello();
      hello();
   </script>
</body></html>
```

Listing 2.26 function_simple.htm File

With functions, a distinction is made between a definition and a call. The browser reads the definition of a function and then knows how it should work, and it's only executed after a call.

A function is often defined in the head of the document. The definition begins with the function keyword, which is followed by the name of the function, for which the same rules apply as for the names of variables (Section 2.1.2). This is followed by parentheses (here still without any content) and then a block of statements, always with curly brackets.

The function is called twice in this example, and it consists of the name of the function followed by the parentheses (here still without any content). You can call the hello() function as often as you like within your program, and each time it's called, the program *jumps* to the function and executes its content.

You can see the output of the program in Figure 2.34.

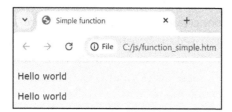

Figure 2.34 Calling Function Twice

2.6.2 Swapping Out Functions

You can swap out the definition of a function that you need frequently and in different programs to an external file. In this way, you can create your own function libraries. Large libraries with many useful functions, such as the jQuery library, are structured in this way.

In Section 1.9, I have already described how you can integrate external files.

Here's the code in the external file:

```
function greetings()
{
    document.write("<p>Function in external file</p>");
}
```

Listing 2.27 external_function.js File

The function is defined in the external file without a script container, and this can be followed by the definition of other functions.

Here's the program that uses the external function:

```
... <head> ...
    <script src="external_function.js"></script>
    <script>
        function hello()
        {
            document.write("<p>Hello world</p>");
        }
    </script>
</head>
<body>
    <script>
        hello();
        greetings();
    </script>
</body></html>
```

Listing 2.28 function_external.htm File

In the first script container, the src attribute is used to reference the external file.

The output of the program is shown in Figure 2.35.

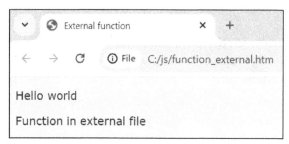

Figure 2.35 Swapping Out Functions

2.6.3 Functions with Parameters

Parameters represent information that's transferred to a function when it's called. The function processes this information and generates results (which may differ depending on the parameters) for the various calls.

You have already used parameters. The alert() method of the window object and the write() method of the document object are methods with parameters, and each of them outputs the string that's transmitted to it as a parameter.

In the program that follows, you can see a separate function to which two strings are passed. The function creates a sentence from this and outputs it.

Here's the program:

```
... <head> ...
    <script>
        function output(country, city)
        {
            document.write("The capital of " + country + " is " + city + "<br>");
        }
    </script>
</head>
<body><p>
    <script>
        const a = "Spain", b = "Madrid";
        output(a, b);
        output("France", "Paris");
    </script></p>
</body></html>
```

Listing 2.29 function_parameters.htm File

A function named output() gets defined, and it has two parameters named country and city, which are separated by commas. The current values of the two parameters are used to compile and output a sentence.

Both variables and values can be transferred to a function. The output() function is called twice: the first time with the a and b variables, and the second time with two strings. Each time it's called, the program jumps to the function, transfers the two parameters in the given order to the country and city variables, and executes the content of the function.

You can see the output of the program in Figure 2.36.

Figure 2.36 Function with Parameters

> **Two Tips**
> - The parameters can be strings, numbers, or truth values. Later, you'll also pass objects or fields using parameters.
> - The number of parameters must match when calling and defining. You'll find additional options in the following sections.

2.6.4 Changing Parameters

You can change parameters within a function. Depending on the type of parameter, this change has an effect on the variables with which the function was called.

A variable of a simple data type (such as a string, number, or truth value) is *called by value*. The variable in the function is a copy, and changing the copy has no effect on the value of the original.

A variable of a nonsimple data type (such as a field) is *called by reference*. In the function, the variable represents a copy of the reference to the original field. The elements of the original field can be changed, but if a different field is assigned to the variable, this action has no effect on the original field. Objects show the same behavior.

The following sample program illustrates this context:

```
... <head> ...
  <script>
    function output(n, t, d, p)
    {
```

```
        document.write(n + ", " + t + ", " + d + ", " + p + "<br>");
    }

    function change(n, t, d, p)
    {
        n = 4711;
        t = "Hello";
        d = false;

        p[0] = "Brad";
        document.write("In the function, element changed: ");
        document.write(p + "<br>");

        p = ["Sarah", "Phil", "Tim"];
        document.write("In the function, field newly created: ");
        document.write(p + "<br>");
    }
    </script>
</head>
<body>
    <p><script>
        const number = 42, tx = "Hello", done = true,
            person = [ "Peter", "Monica", "John" ];
        document.write("Before the call: ");
        output(number, tx, done, person);

        change(number, tx, done, person);
        document.write("After the call: ");
        output(number, tx, done, person);
    </script></p>
</body></html>
```

Listing 2.30 function_parameters_change.htm File

The output() function is used to output the four transferred parameters, which are changed in different ways in the change() function.

Let's first look at the main program in which the functions are called. Four variables are declared, and they are assigned a string, a number, a truth value, and a field. They are then passed twice to the output() function so that you can see the values at the beginning and end of the program. In the simple output of a field, the listed elements are separated by commas.

The change() function is called with the same variables:

- The n variable is a copy of the number variable, and a change of n has no effect on number. The same applies when the variables tx and done are transferred to t and d.

- The p variable is a copy of the person variable, which means that p references the original person field. The change to the first field element is retained, but if a different field is assigned to p, this action only affects p and not the original person field.

This behavior applies regardless of whether the variables are changeable or unchangeable.

The output of the program is shown in Figure 2.37.

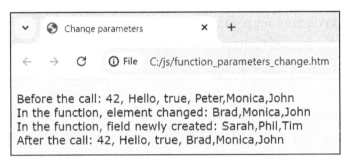

Figure 2.37 Change Parameters

2.6.5   Functions with Return Value

Not only can you send parameters to a function, but you can also receive a result back from a function. This result is referred to as the *return value* of a function, and it can be a string, a number, a truth value, or a field, for example. You can output the return value directly, but you can also save it to use it later in the program.

You have already used return values of functions several times. The isNaN() method returns a truth value, while the Math.random() method returns a random number.

The following program contains a function that requires two numbers as parameters, and it returns the sum of the two numbers:

```
... <head> ...
   <script>
      function sum(a, b)
      {
         const result = a + b;
         return result;
      }
   </script>
</head>
<body>
```

```
  <p><script>
    const x = 3, y = 5;
    const z = sum(x, y);
    document.write("Sum: " + z + "<br>");
    document.write("Sum: " + sum(14, 20/4));
  </script></p>
</body></html>
```

Listing 2.31 function_return.htm File

The sum() function expects two parameters that are added together within the function. The next statement contains the return keyword, and it has two tasks:

- To return from the function from any position
- To return the result of the function to the point at which it's called

The first time the sum() function is called, the current values of x and y are passed to the function. The sum is returned and stored in the z variable, and this is then output.

The document.write() method is called to calculate and output the second sum. Within the call, the sum() function is called again and two numerical values are transferred. The return value is incorporated directly into the output, as shown in Figure 2.38.

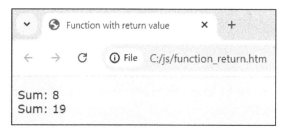

Figure 2.38 Output of Return Value

If the function were to end without a return, or if a return were to occur without a value, the return would have the value undefined.

> **Note**
>
> You'll find the u_function exercise in bonus chapter 1, section 1.7, in the downloadable materials for this book at *www.rheinwerk-computing.com/5875*.

2.6.6   Destructuring Assignment

Functions can only provide a single return value. If you want to use a function to determine and return multiple values or change multiple transferred values, you have two options:

- You can save the values in a field and transfer or return the field.

- You can work with a *destructuring assignment*, as described in this section.

The second technique is used in the following program to swap two values by means of a function:

```
...  <head> ...
   <script>
      function swap(x, y)
      {
         document.write("After the transfer: " + x + " " + y + "<br>");
         const temp = x;
         x = y;
         y = temp;
         document.write("Before the return: " + x + " " + y + "<br>");
         return [x, y];
      }
   </script>
</head>
<body><p>
   <script>
      let a = 7, b = 12;
      document.write("Before the swap: " + a + " " + b + "<br>");
      [a, b] = swap(a, b);
      document.write("After the swap: " + a + " " + b);
   </script></p>
</body></html>
```

Listing 2.32 function_destructuring.htm File

Let's first look at the call of the swap() function. The two variables whose values are to be swapped are passed as parameters, and there's no single return value of the function that's assigned to a single variable.

Instead, a destructuring assignment takes place. The return of the function is passed to multiple variables that are placed like a field within rectangular brackets. If these variables had not already been declared, this could have been done immediately before the rectangular brackets, for example, by using const [c, d] = ...

The output in Figure 2.39 shows that the two values of a and b have been successfully swapped with each other.

In the swap() function, the names of the two parameters are x and y, and they are swapped with each other using a third variable. The function returns the x and y variables, which are enclosed in rectangular brackets like a field. This ensures that x is passed to a and y is passed to b at the call point.

As with calling a function, the number of returned values should match the number of expected values.

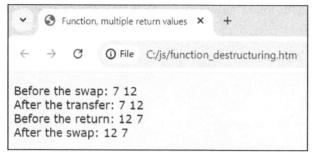

Figure 2.39 Return of Multiple Values

> **Note**
>
> You'll find the u_destructuring exercise in bonus chapter 1, section 1.8, in the down-loadable materials for this book at *www.rheinwerk-computing.com/5875*.

2.6.7   Evaluation Using Short Circuit

In this section, we want to take a look at the *short-circuit* behavior in the evaluation of logic operations:

- If two conditions are linked using the logical OR and the first condition is true, the second condition is no longer examined because the entire condition is already true.

- If two conditions are linked using the logical AND and the first condition is false, the second condition is also no longer examined because the entire condition is already false.

In this example, functions that return a truth value are used within the logical operations. Depending on the return value of the first function, the second function may no longer be called due to the short-circuit behavior.

To understand the following sample program, let's assume we are on a playing field on which the player has the permitted positions from 1 to 10. In relation to certain positions, it should be checked whether certain neighboring positions exist.

```
... <head> ...
  <script>
    function leftNeighbor(p)
    {
      document.write("Function leftNeighbor()<br>");
      if(p > 1)
        return true;
```

```
            else
                return false;
        }

        function rightNeighbor(p)
        {
            document.write("Function rightNeighbor()<br>");
            if(p < 10)
                return true;
            else
                return false;
        }
    </script>
</head>
<body><p>
    <script>
        if(leftNeighbor(5) || rightNeighbor(5))
            document.write("At least one neighbor exists<br>");
        else
            document.write("No neighbor exists<br>");
        document.write("<br>");

        if(leftNeighbor(1) && rightNeighbor(1))
            document.write("Both neighbors exist<br>");
        else
            document.write("Not both neighbors exist<br>");
        document.write("<br>");

        if(leftNeighbor(5) && rightNeighbor(5))
            document.write("Both neighbors exist<br>");
        else
            document.write("Not both neighbors exist<br>");
    </script></p>
</body></html>
```

Listing 2.33 function_short_circuit.htm File

The first logical operation checks whether there's at least one neighbor in relation to position 5. Since there's already a left neighbor, the rightNeighbor() function doesn't get called.

The second logical operation examines whether both neighbors exist in relation to position 1. As the left neighbor doesn't already exist, the rightNeighbor() function doesn't get called here either.

The third logical operation checks whether both neighbors exist in relation to position 5. Since there's a left neighbor, the rightNeighbor() function is also called.

You can see the output of the operations in Figure 2.40.

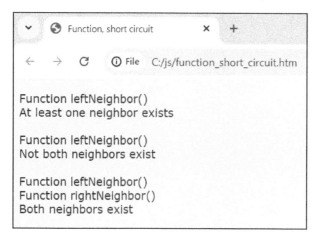

Figure 2.40 Evaluation Using Short Circuit

2.6.8   Any Number of Parameters

In the standard case, the number of parameters when calling a function should match the number of expected parameters. You can use the arguments object to transfer any number of parameters.

The following program shows a function that expects any number of parameters and returns the sum of all parameters:

```
... <head> ...
   <script>
      function sum()
      {
         let result = 0;
         for(let i=0; i<arguments.length; i++)
            result += arguments[i];
         return result;
      }
   </script>
</head>
<body><p>
   <script>
      const x = 12;
      document.write("Sum: " + sum(3 * x, 3.5) + "<br>");
      document.write("Sum: " + sum(-9, 3, x, 5.5) + "<br>");
      document.write("Sum: " + sum());
```

```
   </script></p>
</body></html>
```

Listing 2.34 function_any.htm File

Let's first look at the definition of the function. The parentheses after the name of the function are empty, and all transferred parameters of a function call are not only available within these parentheses but also in the `arguments` object. This object resembles a field, and you can access the first parameter via `arguments[0]`, the second via `arguments [1]`, and so on. The `length` property contains the number of parameters.

To calculate the sum, all parameters are run through using a `for` loop. The sum of the parameters is determined and returned as the return value of the function.

In the program, you'll see three different calls to the function: one with two parameters, one with four parameters, and one without parameters. The sum of the parameters is determined and output, and calling the function without parameters should show that the number of parameters is really arbitrary.

The output of the program is shown in Figure 2.41.

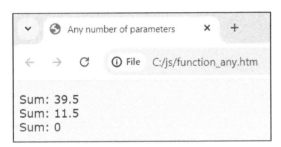

Figure 2.41 Any Number of Parameters

2.6.9   Default Values for Parameters

When defining a function, you can assign default values to individual parameters at the end of the parameters list. This gives you another option for calling functions with different numbers of parameters.

Here's an example:

```
... <head> ...
   <script>
      function add(a, b, c = 0, d = 0)
      {
         const sum = a + b + c + d;
         return sum;
      }
   </script>
```

```
</head>
<body><p>
   <script>
      document.write("Two numbers: " + add(3, 5) + "<br>");
      document.write("Three numbers: " + add(3, 5, 4) + "<br>");
      document.write("Four numbers: " + add(3, 5, 4, 9));
   </script></p>
</body></html>
```

Listing 2.35 function_default.htm File

The `add()` function is used to add two, three, or four values. The last two parameters have the default value 0, so the sum is not distorted. The output of the program is shown in Figure 2.42.

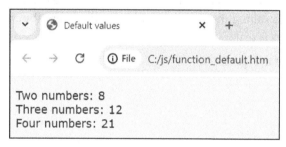

Figure 2.42 Function with Default Values

2.6.10   The Validity of Variables

The validity of a variable, which is declared using `let` or `const`, depends on the location of the declaration:

- If it's declared outside a block of curly brackets, it's valid throughout the entire program.
- If it's declared within a block of curly brackets, it's only valid within the block. The block can also be a function, and you can declare multiple variables with the same name in different blocks. Variables with the same name that have a greater validity are hidden within the block.

A parameter of a function is only valid in the relevant function, and you can't declare a variable with the same name at the top level of the function in question.

In the following program, you can see the connections just described:

```
... <head> ...
   <script>
      function oscar(x)
      {
```

```
      const a = 52;
      const d = 45;
      document.write("In a function: a:" + a + ", b:" + b
        + ", x:" + x + ", d:" + d + "<br>");
      if(true)
      {
        const a = 62;
        document.write("In a block: a:" + a + ", b:" + b
          + ", x:" + x + ", d:" + d + "<br>");
      }
    }
  </script>
</head>
<body><p>
  <script>
    const a = 42;
    const b = 43;
    const c = 44;
    document.write("In the entire program: a:"
      + a + ", b:" + b + ", c:" + c + "<br>");
    oscar(c);
  </script></p>
</body></html>
```

Listing 2.36 function_validity.htm File

The a, b, and c variables are declared in the lower part of the document. They are valid for the entire program, and the c variable gets passed when the oscar() function is called. The x variable in the oscar() function is a copy of c and is only valid within the function.

> **Note**
>
> You could also use the original c variable in the function, but within the function, you should work as little as possible with variables that apply to the entire program. If a variable is required from outside the function, it should be passed as a parameter. This makes it easier to control and more difficult to change accidentally.

Another variable with the name a is declared within the function. It hides the variable of the same name mentioned above and is valid in the entire function, and the d variable is also valid in the entire function.

A variable with the name a is again declared within the block after the if. This variable hides all variables of the same name mentioned above and is valid in the entire if block.

In Figure 2.43, you can see the variables and their values.

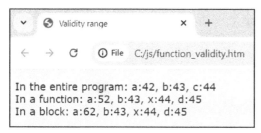

Figure 2.43 Validity of Variables

> **Note**
>
> You should always declare variables as close as possible to where they are used (i.e., within a function or block and just before they are used). This makes them easier to control and harder to change accidentally.

2.6.11   Recursive Functions

Functions can also call other functions. This process is referred to as *nested calls*, in which the program returns to the calling point in each case. Functions can also call themselves, in a process known as *recursion*. A recursive function must contain a branch that terminates the recursion; otherwise, an endless chain of self-calls would result. The most elegant way to solve certain problems is by recursion.

In the following program, a number is halved until it reaches or falls below a certain limit value. You can see the output of the program in Figure 2.44.

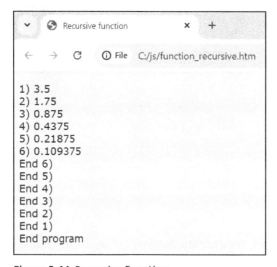

Figure 2.44 Recursive Function

Here's the code of the sample program:

```
... <head> ...
   <script>
      function halve(value, i)
      {
         document.write(i + ") " + value + "<br>");
         value /= 2;
         i++;
         if(value > 0.1)
            halve(value, i);
         document.write("End " + (i-1) + ")<br>");
      }
   </script>
</head>
<body><p>
   <script>
      const value = 3.5, i = 1;
      halve(value, i);
      document.write("End program");
   </script></p>
</body></html>
```

Listing 2.37 function_recursive.htm File

The halve() function is called with the initial value of 3.5. The value gets output and then halved, and if the value is greater than the limit value of 0.1, the function calls itself again. If the limit value is reached or fallen short of, the recursion ends. The i variable is used here as a counter for the number of calls.

In this example, the end of the recursion is reached with the sixth call and the function ends for the first time. The sequence returns to the end of the fifth run, then to the end of the fourth run, and so on until the end of the first run. Finally, the end of the program is reached.

You can find further examples of recursion in this book, for example, in Chapter 5, Section 5.3.

2.6.12 Anonymous Functions

Up to this point, we have been using *named functions* (i.e., functions that have a name). However, there are also *anonymous functions* (i.e., functions without a name). They look a little unusual at first, but in certain situations, they make it possible to significantly shorten the program code. For example, anonymous functions are used for handling events (see Chapter 4) or for using the extensive jQuery library (see Chapter 11).

In addition, you can work not only with references to fields or objects but also with references to functions. A reference to a function can be assigned to a variable, and the reference to a function can also be passed as a parameter.

Here's a sample program:

```
... <head> ...
   <script>
      const greeting = function()
      {
         document.write("Hello world<br>");
      }

      const sum = function(a,b)
      {
         const s = a + b;
         return s;
      }

      function oscar(welcome)
      {
         welcome();
      }
   </script>
</head>
<body><p>
   <script>
      greeting();
      oscar(greeting);

      const result = sum(35, 7);
      document.write(result + "<br>");

      const add = sum;
      document.write(add(45, 17) + "<br>");

      document.write((function() { return "Good morning<br>"; })());
      document.write((function(a, b) { return a * b; })(12, 6));
   </script></p>
</body></html>
```

Listing 2.38 function_anonymous.htm File

The first anonymous function outputs the text Hello world, and it's defined without parameters. The reference to this function is assigned to the greeting variable.

In the second anonymous function, the two transferred parameters are added together, and the result is delivered as a return value. The reference to the function is assigned to the sum variable.

Each of the two anonymous functions is called using the name of the variable followed by parentheses, as with a named function: greeting() or sum(35, 7).

You can transfer the reference to a function to another function. This is what happened here when the oscar() function was called. Within the oscar() function, the welcome name references the first anonymous function, and for this reason, the welcome() call leads to the same result.

The value of a variable that references a function can be assigned to another variable. Here the variable sum was assigned to the variable add, and you can then start the second anonymous function using the add(...) call.

At the end of the program, you can see two anonymous functions that are not assigned to any variables but are called directly:

- The first function has no parameters and returns the text "Good morning". The returned text is output using document.write().

- The second anonymous function has the a and b parameters and returns the product of these two parameters. Here, it's called with the values 12 and 6, and again, the return result is output using document.write().

To call an anonymous function, you must enclose its entire definition in parentheses. This is followed by the parentheses for the parameters, if available. Always keep track of the many levels of parentheses. In Figure 2.45 you can see the result of the program.

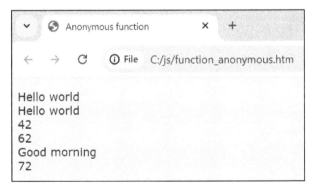

Figure 2.45 Anonymous Functions

2.6.13   Callback Functions

The reference to a function can be transferred to another function for internal use. In this context, the transferred function is referred to as the *callback function*. You can transfer both named functions and anonymous functions as callback functions. Callback functions are also used to handle events (see Chapter 4).

In the following program, the output() function is called twice. It's used to output a regular sequence of values. A mathematical calculation is then carried out for each of these values, the result of which is displayed next to the value.

The mathematical calculation can be freely selected. It's defined in a function, and a reference to this function is transferred to the output() function when it's called:

- The first call passes a reference to the named function square(), which returns the square of the transferred value.
- The second call transfers an anonymous function that returns the reciprocal of the transferred value.

Here's the program:

```
... <head> ...
   <script>
      function square(x)
      {
         return x * x;
      }

      function output(from, to, step, fct)
      {
         for(let p = from; p < to + step / 2.0; p += step)
            document.write(p + " " + fct(p) + "<br>");
         document.write("<br>");
      }
   </script>
</head>
<body><p>
   <script>
      output(5.0, 15.0, 2.5, square);
      output(3.0, 7.0, 1.0, function(p) { return 1 / p; });
   </script></p>
</body></html>
```

Listing 2.39 function_callback.htm File

The first call calculates the squares of the numbers from 5 to 15 in steps of 2.5, and the second call calculates the reciprocal values of the numbers from 3 to 7 in steps of 1.

In contrast to the program from Section 2.6.12, the anonymous function doesn't have to be placed in parentheses as it's not called here. Only a reference to the function is transferred.

The output of the program is shown in Figure 2.46.

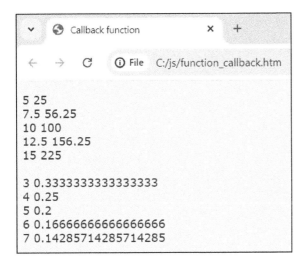

Figure 2.46 Callback Function

Chapter 3
Custom Objects

In this chapter, you'll learn about creating your own objects and gain a better understanding of the structure of existing objects.

In JavaScript, you program in an object-oriented way. There's a whole range of predefined objects:

- You already know about the window, document, and Math objects.
- In Chapter 4, Section 4.8, you'll use the location object.
- Chapter 6 describes other predefined objects.

However, you can also create your own objects, as I will describe in this chapter. This allows you to bundle and access thematically related data, which makes it easier for you to understand and handle the numerous predefined objects. If you need special options in your applications, you can extend both your own objects and predefined objects by means of *inheritance*.

Traditionally, custom objects are created in JavaScript using prototypes and constructor functions. Since the introduction of the ECMAScript 2015 standard, there has been a new standard notation in which classes, constructors, properties, and methods are used, as in other object-oriented languages. All modern browsers use this notation, and in this book, only this kind of notation is used.

The *prototype* term will only be discussed in individual, separate contexts.

3.1 Objects and Properties

This section describes how you can define a class with the name Car and the properties color and speed. The class that is first defined will be used in the actual program, and then two objects of that class will be created and output. Different objects of the same class are related to each other, and they have the same properties but with different values.

Here's the program:

```
... <head> ...
   <script>
```

```
    class Car
    {
       constructor(c, s)
       {
          this.color = c;
          this.speed = s;
       }
    }
  </script>
</head>
<body>
  <p><script>
     const dodge = new Car("Red", 50);
     document.write("Color: " + dodge.color
        + ", Speed: " + dodge.speed + "<br>");

     dodge.speed = 75;
     document.write("Color: " + dodge.color
        + ", Speed: " + dodge.speed + "<br>");

     let renault = new Car("Yellow", 65);
     renault = new Car("Blue", 85);
     document.write("Color: " + renault.color
        + ", Speed: " + renault.speed);
  </script></p>
</body></html>
```

Listing 3.1 obj_property.htm File

The class keyword introduces the definition of the class. Note that the name of the class is not followed by parentheses, as would be the case with a function.

The constructor method is used to create an object of the class. It has the fixed name constructor, and its parameters correspond to the properties of an object of the class.

Two properties are defined within the constructor method, each with the this keyword, the *dot operator*, and a name. The two properties are assigned the values that are transferred as parameters when an object of the Car class gets created. The this keyword references *this object* (i.e., the current object), and it ensures that a property of the current object is accessed or a method is executed for the current object.

The defined class is used in the actual program. The unchangeable dodge variable is declared, and it's assigned a reference to a new object of the Car class using the new keyword. Two values are transferred as the initial values of the two properties, and you can then access the properties of the object using dot notation. The values can be output or changed.

Then, the `renault` variable is declared, and an object of the `Car` class is assigned to this variable twice in succession. These objects also have the `color` and `speed` properties, but the property values of the objects differ.

When declaring variables that reference objects, the situation is the same for objects as for fields: If an unchangeable variable references an object, only the reference is unchangeable. You can change the individual elements of the object, but you can't assign another object to the variable.

The output of the program is shown in Figure 3.1.

Figure 3.1 Objects and Properties

3.2 Methods

A *method* is a function that can only be called for objects of the class in which it was defined. For example, the `write()` method shows only the familiar behavior if it's called using the dot operator for the `document` object. Methods are often used to change the properties of an object of the relevant class, and the properties and methods of a class are referred to together as *members* of this class.

In the following example, the `Car` class from Section 3.1 gets extended. The `paint()` and `accelerate()` methods are defined and used to change the `color` and `speed` properties. In addition, a special method with the name `toString()` is defined; it enables the simple output of an object, similar to the output of a variable.

Here's the program:

```
... <head> ...
   <script>
      class Car
      {
         constructor(c, s)
         {
            this.color = c;
            this.speed = s;
         }
```

```
        accelerate(value)
        {
           this.speed += value;
        }

        paint(c)
        {
           this.color = c;
        }

        toString()
        {
           return "Color: " + this.color
              + ", speed: " + this.speed;
        }
      }
   </script>
</head>
<body><p>
   <script>
      const dodge = new Car("Red", 50);
      document.write("Color: " + dodge.color
         + ", Speed: " + dodge.speed + "<br>");

      dodge.accelerate(35);
      dodge.paint("Blue");
      document.write(dodge);
   </script></p>
</body></html>
```

Listing 3.2 obj_method.htm File

The accelerate() method is called using the dot operator for the dodge object. This method expects a numerical value that is used to change the value of the speed property, while the paint() method expects a character string that represents the new value of the color property.

The special toString() method is called automatically if the object itself is to be output when the document.write() method is called. The toString() method returns a string containing the values of all the object's properties (see Figure 3.2).

Note

You can find the u_class exercise in bonus chapter 1, section 1.9, in the downloadable materials for this book at *www.rheinwerk-computing.com/5875.*

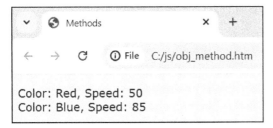

Figure 3.2 Using Methods

3.3 Private Members

In the previous programs in this chapter, you were able to access the properties and methods of objects directly from the actual program. The members in those programs are therefore publicly accessible, or *public* for short.

In accordance with the object-oriented principle of data encapsulation, however, the properties of an object should remain as protected as possible. There should only be controlled access to these properties using methods, and certain methods that are only used for the internal processes of a class should also be protected. Such protected members are called *private members*.

Since ECMAScript 2022, you have been able to declare private members using the hash character, #. The following is a sample program in which a private property and a private method are declared:

```
... <head> ...
  <script>
    class Car
    {
      #color;

      constructor(c, s)
      {
        this.#color = c;
        this.speed = s;
      }

      accelerate(value)
      {
        this.speed += value;
      }

      #paint(c)
      {
```

```
        if(c == "Red" || c == "Yellow" || c == "Blue")
            this.#color = c;
    }

    change(c, value)
    {
        this.#paint(c);
        this.accelerate(value);
    }

    toString()
    {
      return "Color: " + this.#color
          + ", speed: " + this.speed;
    }
  }
}
    </script>
</head>
<body>
    <p><script>
      const dodge = new Car("Red", 50);
      document.write(dodge + "<br>");
      // document.write("Color: " + dodge.#color + "<br>");

      dodge.change("Green", 25);
      document.write(dodge + "<br>");
      dodge.change("Blue", 10);
      document.write(dodge + "<br>");
      // dodge.#paint("Yellow");
    </script></p>
</body></html>
```

Listing 3.3 obj_private.htm File

A private property is declared outside the methods, usually at the beginning of the class definition. Here, this is done using the #color; statement. Within the class definition, this property is accessed using this.#color. Access outside the class definition is no longer possible.

To define a private method, # is noted before the method's name, as you can see in the #paint(c) method header. Within the class definition, this method is accessed using this.#paint(...). Access outside the class definition is no longer possible.

In the paint() method, care is taken to ensure that only the permitted colors red, yellow or blue can be used. In this way, the private property color is protected. You can see the result of the program in Figure 3.3.

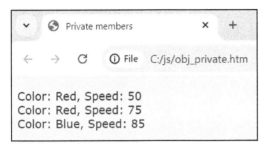

Figure 3.3 Access to Private Members

3.4 Setters and Getters

Since ECMA 2022, you can also use setters and getters (so-called *accessors*) for controlled access to properties. A *setter* is used for write access using the set keyword, while a *getter* is used for read access using the get keyword.

The following is a sample program in which accessors are defined for a private property:

```
... <head> ...
  <script>
    class Car
    {
      #speed;

      constructor(c, s)
      {
        this.color = c;
        this.setSpeed = s;
      }

      get getSpeed()
      {
        return this.#speed;
      }

      set setSpeed(s)
      {
        if(s < 0)        s = 0;
        else if(s > 100) s = 100;
```

```
            this.#speed = s;
        }

        accelerate(value)
        {
            this.setSpeed = this.getSpeed + value;
        }

        toString()
        {
            return "Color: " + this.color
                + ", Speed: " + this.getSpeed;
        }
    }
</script>
</head>
<body>
    <p><script>
        const dodge = new Car("Red", 150);
        document.write(dodge + "<br>");
        dodge.accelerate(20);
        document.write(dodge + "<br>");
        dodge.accelerate(-120);
        document.write(dodge + "<br>");
        dodge.accelerate(30);
        document.write(dodge + "<br>");
    </script></p>
</body></html>
```

Listing 3.4 obj_accessor.htm File

In the program, the value of the private speed property can only be changed using the accelerate() method. In this method, the current value of the speed property is first determined using the getSpeed getter, and the getter returns the value like a method using return.

A new value for the speed property is calculated on the right-hand side of the assignment. The assignment itself leads to a call of the setSpeed setter, and the new value is transferred as a parameter, as with a method. The setter ensures that the new value for the speed is only within the permitted range between 0 and 100, and it's then assigned to the private speed property.

The setter is also used in the constructor of the class to limit the value of the property. The output of the program looks as shown in Figure 3.4.

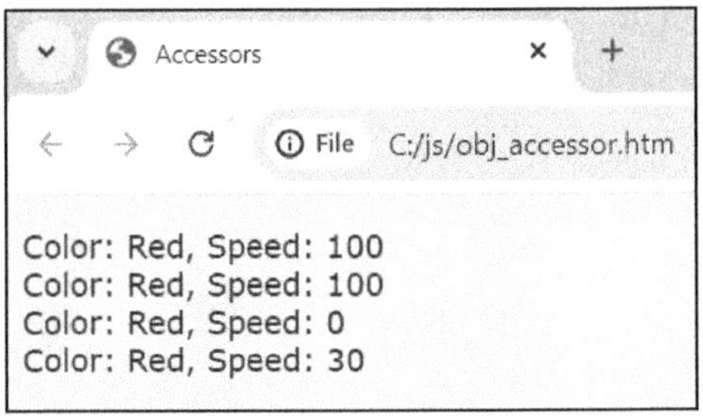

Figure 3.4 Control with Accessors

3.5   Static Members

Different objects of the same class usually have different values for their properties. In addition to these object-related properties, you can define *static* properties, which relate to the class as a whole and are available to all objects with the same value. You can also define static methods, which also refer to an entire class and are called for this class.

The following is a sample program in which a static property is declared in a class using the static keyword and a static method is defined:

```
... <head> ...
  <script>
    class Car
    {
      static quantity = 0;

      constructor(c, s)
      {
        this.color = c;
        this.speed = s;

        Car.quantity = Car.quantity + 1;
        this.number = Car.quantity;
      }

      toString()
      {
        return "Color: " + this.color + ", Speed: "
            + this.speed + ", Number: " + this.number;
      }

      static output_quantity()
      {
```

```
                    document.write("Quantity: " + Car.quantity + "<br>");
                }
            }
    </script>
</head>
<body>
    <p><script>
        const dodge = new Car("Red", 50);
        document.write(dodge + "<br>");
        const renault = new Car("Yellow", 65);
        document.write(renault + "<br>");

        document.write("Quantity: " + Car.quantity + "<br>");

        Car.output_quantity();
    </script></p>
</body></html>
```

Listing 3.5 obj_static.htm File

A static property is declared outside the methods, usually at the beginning of the class definition, and it's given a start value. Here, we have read and write access to it in the constructor.

The static quantity property is used to determine the number of objects created in the class. It also gives the object-related number property a separate value for each object, which represents its sequential number in the order in which it was created.

The static quantity property must be accessed within the methods and in the program via the name of the class (i.e., via Car.quantity).

The static output_quantity() method outputs a line that specifies the number of objects, and the method must be called in the program or within other methods using the name of the class (i.e., Car.output_quantity()).

The output of the program is shown in Figure 3.5.

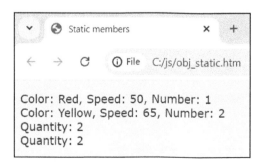

Figure 3.5 Use of Static Members

3.6   Static Blocks

Since ECMAScript 2022, you've been able to use static blocks for the one-time initialization of a class. They are usually also noted at the beginning of the class definition, and static properties can be defined in such blocks of statements. Such blocks could also contain other statements to be executed.

In the following program excerpt, you'll only see those lines that differ from the *obj_static.htm* program:

```
... <head> ...
   <script>
      class Car
      {
         static
         {
            Car.quantity = 0;
         }

         constructor(c, s)
         { ...
```

Listing 3.6 obj_static_block.htm File Excerpt

The static block is introduced using the static keyword. Variables that are declared in it are static, and as the variables are located within the block, they must be accessed via the name of the class. The output of the program corresponds to that of the *obj_static.htm* program, as you can see in Figure 3.5.

3.7   Reference to Nothing

In connection with objects, a given variable can reference nothing instead of an object. This can be the case after the given variable has been assigned the value null, but it can also happen if the variable is supposed to reference a DOM element (see Chapter 5) that doesn't exist in the document.

A variable that references nothing has the value null, and a *reference to nothing* often leads to the program being aborted. There are two operators that can prevent this from happening:

- The *null coalescing* operator, ??
- The *optional dot* operator, ?.

The following program shows examples of using these operators:

```
... <head> ...
   <script>
      class Car
      {
         constructor(c, s)
         {
            this.color = c;
            this.speed = s;
         }

         toString()
         {
            return "Color: " + this.color
               + ", speed: " + this.speed;
         }
      }
   </script>
</head>
<body><p>
   <script>
      let dodge = new Car("Red", 50);
      document.write((dodge ?? "Object does not exist") + "<br>");
      document.write("Color: " + dodge?.color + "<br>");

      dodge = null;
      document.write((dodge ?? "Object does not exist") + "<br>");
      // document.write("Color: " + dodge.color + "<br>");
      document.write("Color: " + dodge?.color + "<br>");
   </script></p>
</body></html>
```

Listing 3.7 obj_null.htm File

When you use the null coalescing operator, ??, the left operand gets checked. If its value is null, the right operand will be returned; otherwise, the left operand will be returned.

Then, since the object that the dodge variable references exists, the object gets output. The reference is later assigned the null value in the program for demonstration purposes, which means that dodge is a reference to nothing, which causes the "Object does not exist" text to be displayed.

The dot operator, ., is used to access the properties and methods of an object. If the object doesn't exist, the program will be aborted. For this reason, the penultimate statement in the program is commented out.

The optional dot operator, ?., is also used to access the properties and methods of an object. If the object doesn't exist, the undefined value will be returned and the program will not abort.

The output of the program is shown in Figure 3.6.

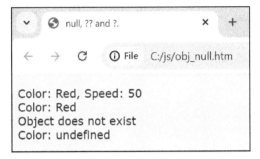

Figure 3.6 Reference to Nothing

3.8   Object in an Object

The property of an object of one class can in turn be an object of another class. To clarify this relationship, we want to extend the Car class by the drive property of the Engine class.

Here's the sample program:

```
... <head> ...
  <script>
    class Engine
    {
      constructor(p, c, f)
      {
        this.power = p;
        this.cylinders = c;
        this.fuel = f;
      }

      tune(x)
      {
        this.power += x;
      }

      toString()
      {
        return "Power: " + this.power + ", Cylinders: "
          + this.cylinders + ", fuel: " + this.fuel;
```

```
            }
        }

        class Car
        {
            constructor(c, s, d)
            {
                this.color = c;
                this.speed = s;
                this.drive = d;
            }

            toString()
            {
                return "Color: " + this.color + ", Speed: "
                    + this.speed + ", Drive: " + this.drive;
            }
        }
    </script>
</head>
<body><p>
    <script>
        const dodge = new Car("Red", 50,
            new Engine(60, 4, "Diesel"));
        dodge.drive.tune(10);
        document.write(dodge + "<br>");

        dodge.drive.power = 80;
        dodge.drive.cylinders = 6;
        dodge.drive.fuel = "Gasoline";
        document.write(dodge);
    </script></p>
</body></html>
```

Listing 3.8 obj_in_object.htm File

An object of the Engine class has the power, cylinders, and fuel properties. In addition, there's the tune() method for changing the performance and the toString() method for outputting the three property values.

The drive property has been added to the Car class, and the value of this property is also output using the toString() method.

An object of the Car class is created in the program, and the third parameter is an object of the Engine class, which is also created using new.

The first dot after the name of the dodge object addresses the drive property of the Car object, and the second dot leads to the subproperty of the Engine object.

The toString() method of the Car class is called to output a Car object, and the method internally calls the method of the same name of the Engine class using this.drive. The entire string is structured in this way, as you can see in Figure 3.7.

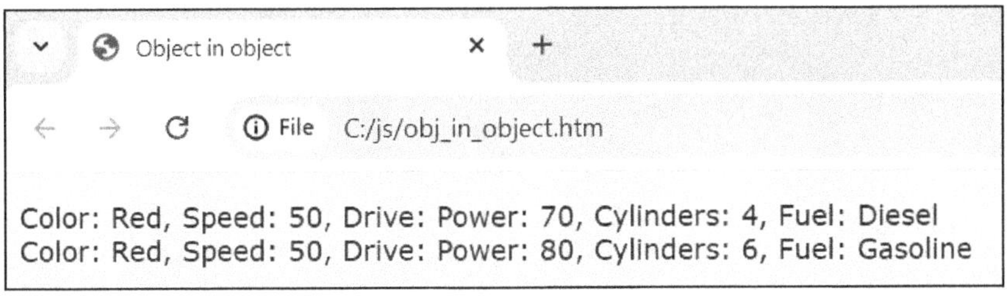

Figure 3.7 Engine Object in Car Object

3.9   Inheritance

If you need a class that has properties that have already been defined in another class, you can use the principle of *inheritance*. In this context, you create a derived class on the basis of a base class and add additional properties.

In the following example, the derived Truck class is created on the basis of the Car base class. An object of the Truck class is supposed to have the additional payload property and the additional load() method, and each of the two classes has its own definition of the toString() method.

Here's the program:

```
... <head> ...
   <script>
      class Car
      {
         constructor(c, s)
         {
            this.color = c;
            this.speed = s;
         }

         accelerate(value)
         {
            this.speed += value;
         }

         toString()
```

```
      {
         return "Color: " + this.color
            + ", Speed: " + this.speed;
      }
   }

   class Truck extends Car
   {
      constructor(c, s, p)
      {
         super(c, s);
         this.payload = p;
      }

      load(value)
      {
         this.payload += value;
      }

      toString()
      {
         return super.toString() + ", Payload: " + this.payload;
      }
   }
</script>
</head>
<body><p>
   <script>
      const kenworth = new Truck("Orange", 30, 15);
      document.write(kenworth + "<br>");

      kenworth.accelerate(50);
      kenworth.load(25);
      document.write(kenworth);
   </script></p>
</body></html>
```

Listing 3.9 obj_inheritance.htm File

The constructor of the Car base class expects two parameters for the initial values of the color and speed properties. The constructor of the derived Truck class expects three parameters, which are the initial values for the color and speed properties of the Car base class and for the payload property of the derived Truck class.

The super() method calls the specific constructor of the base class within a constructor, and in this way, the first two parameters are passed on to the constructor of the base class. The third parameter is assigned to the payload property, and in this way, all properties of the Truck class object are given an initial value.

There's also the super reference (without parentheses), which is used to access the properties and methods of the base class. In the toString() method of the derived Truck class, the method of the same name of the Car base class is called: super.toString().

The program creates and outputs an object of the Truck class with three initial values. It's changed by calling the accelerate() method of the base class and by calling the load() method of the derived class. Then, it's output again, as shown in Figure 3.8.

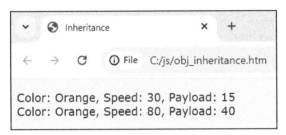

Figure 3.8 Inheritance

> **Note**
>
> Regarding the topic of inheritance, you can find the u_inheritance exercise in bonus chapter 1, section 1.10, in the downloadable materials for this book at *www.rheinwerk-publishing.com/5875*.

3.10 Operations with Objects

In this section, you'll see a sample program in which a number of useful operations and checks are carried out on a class, its properties and methods, and some objects of the class. The class is defined in the head of the document as follows:

```
... <head> ...
  <script>
    class Car
    {
      constructor(c, s)
      {
        this.color = c;
        this.speed = s;
      }
```

```
      accelerate(value)
      {
         this.speed += value;
      }

      paint(c)
      {
         this.color = c;
      }

      toString()
      {
         return "Color: " + this.color
            + ", Speed: " + this.speed;
      }
   }
   </script>
</head>
```

Listing 3.10 obj_operation.htm File, Part 1 of 8

3.10.1   Access Operators

You have two options for accessing the properties of an object. First, you can use the dot operator, .; and second, you can use square brackets, [ ]. You can see this in the following program section:

```
<body><p>
   <script>
      const dodge = new Car("Red", 50);
      document.write("Notation with dot: " + dodge.color + " "
         + dodge.speed + "<br>");
      document.write("Notation with [ and ]: " + dodge["color"]
         + " " + dodge["speed"] + "<br>");
      document.write("Notation with toString(): "
         + dodge + "<br><br>");
```

Listing 3.11 obj_operation.htm File, Part 2 of 8

Once an object has been created, its properties are output multiple times. Within the square brackets, the name of the property is written like a string within double quotation marks. You can compose this string from individual parts, also using variables, and this kind of notation makes the creation of programs even more flexible. You can see the output, among others, in Figure 3.9.

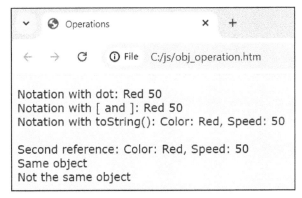

Notation with dot: Red 50
Notation with [and]: Red 50
Notation with toString(): Color: Red, Speed: 50

Second reference: Color: Red, Speed: 50
Same object
Not the same object

Figure 3.9 Access Options and Comparisons

3.10.2   Creating and Comparing References to Objects

If you assign a variable that references an object to another variable, you don't create a new object or a copy of the original object but only a second reference to the same object. You can then access the same object via both variables (see also Figure 3.9).

The two comparison operators, == and !=, can be used to check whether two variables reference the same object.

Variables are compared with each other in the following part of the program, in the following ways:

- First, there are two variables that reference the same object.

- Second, there are two variables that reference two different objects with the same properties and property values.

```
const renault = dodge;
document.write("Second reference: " + renault + "<br>");
if(renault == dodge)
    document.write("Same object<br>");
const simca = new Car("Red", 50);
if(simca != dodge)
    document.write("Not the same object<br><br>");
```

Listing 3.12 obj_operation.htm File, Part 3 of 8

You can also see this output in Figure 3.9.

Regarding the copying of objects, please refer to Section 3.11.

3.10.3   Checking Instances

Objects of a class are referred to as *instances* of a class, and the process of creating an object is called *instantiation*. The instanceof operator checks whether an object represents the instance of a specific class (or its base class):

```
if(dodge instanceof Car)
    document.write("instanceof:<br>Object is an instance of the class<br><br>");
```

Listing 3.13 obj_operation.htm File, Part 4 of 8

You can see the output in Figure 3.10.

```
instanceof:
Object is an instance of the class

typeof:
Car: function
dodge: object
color: string
speed: number
accelerate: function
paint: function
```

Figure 3.10 instanceof and typeof Operators

3.10.4   Determining a Type

The typeof operator returns the type of a variable (see also Chapter 2, Section 2.3.6), the methods of a class have the *function* type, and the objects of a class have the *object* type. Classes in JavaScript are based on prototypes and constructor functions, and for this reason, the classes themselves also have the function type, as follows:

```
document.write("typeof:<br>"
    + "Car: " + typeof Car + "<br>"
    + "dodge: " + typeof dodge + "<br>"
    + "color: " + typeof dodge.color + "<br>"
    + "speed: " + typeof dodge.speed + "<br>"
    + "accelerate: " + typeof dodge.accelerate + "<br>"
    + "paint: " + typeof dodge.paint + "<br><br>");
```

Listing 3.14 obj_operation.htm File, Part 5 of 8

In Figure 3.10, you can see that the Car class and the accelerate() and paint() methods have the function type while the dodge object has the object type. The two properties are simple variables of the *string* or *number* type.

3.10.5   Checking a Member

You can use the in operator to check whether a class has a specific member (i.e., a specific property or method):

```
document.write("in:<br>");
if("color" in dodge)
    document.write("color is member<br>");
if("accelerate" in dodge)
    document.write("accelerate is member<br>");
if("paint" in dodge)
    document.write("paint is member<br>");
if(!("power" in dodge))
    document.write("power is no member<br><br>");
```

Listing 3.15 obj_operation.htm File, Part 6 of 8

The check shows that the color, accelerate, and paint strings are members of the class of the dodge object, but power is not, as shown in Figure 3.11.

```
in:
color is member
accelerate is member
paint is member
power is no member

Color: Red, Speed: 60
```

Figure 3.11 in Operator, Objects and Functions

3.10.6   Objects and Functions

You can pass a variable that represents a reference to an object as a parameter to a function. Within the function, you can change the object or call methods of the object, and you can also return a variable that represents a reference to an object as a return value from a function and assign it to another variable:

```
function change(x)
{
    x.accelerate(10);
    return x;
}
const lada = change(dodge);
document.write(lada + "<br><br>");
```

Listing 3.16 obj_operation.htm File, Part 7 of 8

The dodge variable, the x parameter, and the lada variable reference the same object. You can see the output in Figure 3.11 as well.

3.10.7   Deleting Properties

You can delete individual properties of an object using the delete operator. The properties of other objects of the same class are not affected by this, but this process has the side effect that the objects are no longer similar.

Finally, here's the last part of the program:

```
        delete dodge.speed;
        document.write("Values of members:<br>"
            + "color: " + dodge.color + "<br>"
            + "speed: " + dodge.speed + "<br>"
            + "paint: " + dodge.paint + "<br>"
            + "power: " + dodge.power);
    </script></p>
</body></html>
```

Listing 3.17 obj_operation.htm File, Part 8 of 8

The speed property gets deleted, and after that, the values of the color, speed, paint, and power properties are output. The speed property no longer exists, while the power property never existed. You can see the resulting output in Figure 3.12.

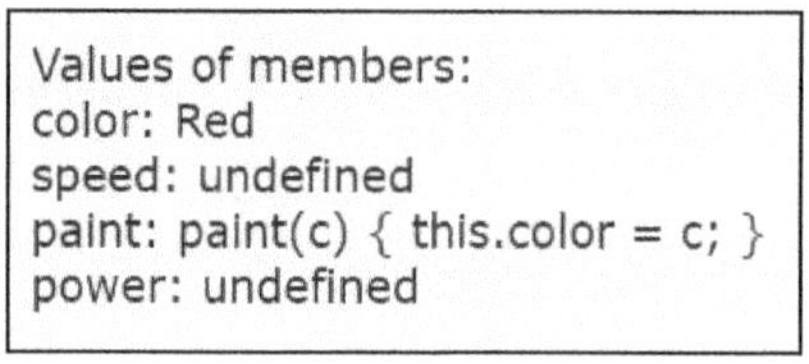

```
Values of members:
color: Red
speed: undefined
paint: paint(c) { this.color = c; }
power: undefined
```

Figure 3.12 Property Values after a Deletion Process

3.11   Copying Objects

To copy an object, it's not sufficient to create a copy of the variable that references the object (see also Section 3.10.2). Instead, a new object, whose properties are assigned the values of the original object, must be created.

The class in the following program contains a method for copying objects of the class:

```
... <head> ...
    <script>
        class Car
        {
```

```
            constructor(c, s)
            {
               this.color = c;
               this.speed = s;
            }

            accelerate(value)
            {
               this.speed += value;
            }

            toString()
            {
               return "Color: " + this.color
                  + ", Speed: " + this.speed;
            }

            copy()
            {
               const c = new Car(this.color, this.speed);
               return c;
            }
         }
      </script>
   </head>
   <body><p>
      <script>
         const dodge = new Car("Red", 50);
         document.write(dodge + "<br>");

         const renault = dodge.copy();

         dodge.accelerate(25);
         document.write(dodge + "<br>");
         document.write(renault + "<br>");
      </script></p>
   </body></html>
```

Listing 3.18 obj_copy_flat.htm File

A new object of the Car class is created in the copy() method, and the values of the two properties of the current object are transferred to the constructor. This means that the new object is a copy of the current object. The copy() method returns a reference to this new object.

The copy() method for the dodge object is called in the program, and the result of the method (i.e., the reference to the new object) is assigned to the renault variable. Two objects whose properties can have different values are then available.

The output of the program looks as shown in Figure 3.13.

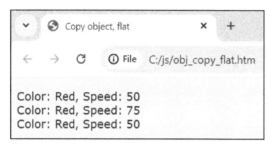

Figure 3.13 Original and Copy

The properties of the class from the preceding example can be copied using a simple assignment, as they have a simple data type, such as string, number, or truth value. In this case, a so-called *flat copy* is sufficient.

However, if a property has a nonsimple data type, such as a field or an object, a so-called *deep copy* must be made. Making a deep copy involves copying the values of all elements of a field to a new field or copying the values of the properties of an object to a new object.

The Car and Engine classes that you already know from Section 3.9 are used next, and the drive property of the Car class is an object of the Engine class. The following is an excerpt from a program in which an object of the Car class is copied using a total of two copy methods:

```
... <head> ...
  <script>
    class Engine
    {
      constructor(p, c, f)
      {
        this.power = p;
        this.cylinders = c;
        this.fuel = f;
      }

      tune(x)
      {
        this.power += x;
      }
```

```
      toString()
      {
         return "Power: " + this.power + ", Cylinders: "
            + this.cylinders + ", Fuel: " + this.fuel;
      }

      copy()
      {
         const e = new Engine(this.power, this.cylinders, this.fuel);
         return e;
      }
   }

   class Car
   {
      constructor(c, s, d)
      {
         this.color = c;
         this.speed = s;
         this.drive = d;
      }

      toString()
      {
         return "Color: " + this.color + ", Speed: "
            + this.speed + ", Drive: " + this.drive;
      }

      copy()
      {
         const c = new Car(this.color,
            this.speed, this.drive.copy());
         return c;
      }
   }
   </script>
</head>
<body><p>
   <script>
      const dodge = new Car("Red", 50, new Engine(60, 4, "Diesel"));
      document.write(dodge + "<br>");

      const renault = dodge.copy();
```

```
        dodge.drive.tune(15);
        document.write(dodge + "<br>");
        document.write(renault + "<br>");
    </script></p>
</body></html>
```

Listing 3.19 obj_copy_deep.htm File Excerpt

When an object of the Car class gets copied, the copy() method of this class is called. This creates a new object of the Car class, and the values of the color and speed properties of the original object are assigned.

For the third property (drive), the process is somewhat more complex. First, the copy() method of the Engine class is called for this property, and this creates a new object of the Engine class. The values of the three properties of the original object of the Engine class are assigned, and a reference to the newly created object is returned to the copy() method of the Car class and assigned to the drive property of the Car class.

You can see the result of the program in Figure 3.14.

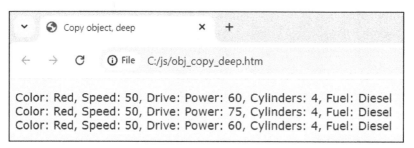

Figure 3.14 Original and Deep Copy

Chapter 4

Forms and Events

In this chapter, you'll learn how to check forms and handle events.

Forms enable interactions similar to those you're used to from other applications on the computer. You can make entries and trigger processing, and the program then returns a result.

Forms are also used to transmit data to a web server. Before you send them, their contents can be checked for validity by JavaScript. This avoids unnecessary network traffic.

JavaScript also allows you to write programs that can respond to *events* in the browser. In this way, you can interact with your programs beyond the simple return values of the prompt() and confirm() functions. These events can involve the click of a button, the selection of an entry from a list, a specific action using the mouse, the submission of a form, and much more.

If you want your program to respond to an event, you need to develop JavaScript program code with statements that must be executed when the event occurs. An *event handler* is used to handle the event; it creates a link between the part of the document where the event is triggered, the event itself, and the JavaScript program code.

4.1 First Form and First Event

In this section, you'll learn about some typical elements that are required to handle an event.

The following example shows a form with the input field **Input** and the **Click** button. After entering a text in the input field, you must click the button (see Figure 4.1), and then a dialog box with information appears (see Figure 4.2).

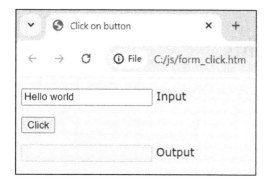

Figure 4.1 Form with Input Field and Button

Figure 4.2 Response to Event

You can also use input fields to subsequently output information within documents that have already been mapped in their entirety. In the example, the same information appears in the **Output** field as in the dialog box (see Figure 4.3). You can use the `disabled` attribute to ensure that no entries can be made in this field.

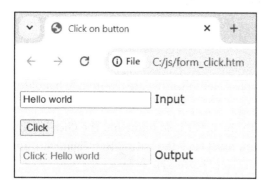

Figure 4.3 Field for Output

Here's the program:

```
... <head> ...
  <script>
    function clicked()
    {
      const inputField = document.getElementById("idInput");
```

```
         const input = inputField.value;
         alert("Click: " + input);
         const outputField = document.getElementById("idOutput");
         outputField.value = "Click: " + input;
      }
   </script>
</head>
<body>
   <form>
      <p><input id="idInput"> Input</p>
      <p><input id="idButton" type="button" value="Click"></p>
      <p><input id="idOutput" disabled> Output</p>
   </form>
   <script>
      const bu = document.getElementById("idButton");
      bu.addEventListener("click", clicked);
   </script>
</body></html>
```

Listing 4.1 form_click.htm File

The elements of a form are in a form container, and a simple element of the input type corresponds to a single-line input field. If the button value is specified for the type attribute of an input element, a button is displayed, and the value of the value attribute corresponds to the text on the button.

The value of the id attribute serves as a unique *identifier* (ID) of the element in the document. This ID is required to access the element and connect to the JavaScript program code.

In the lower script container, the getElementById() method of the document object is called. It expects the ID of an element as a parameter and returns a reference to an object that represents this element. In the example, this reference is saved in the bu variable, which thus references the button.

An *event listener* detects an event and responds to it, and the addEventListener() method of the element object links an element of the document with a specific event and with a reference to a named or anonymous function. In the present case, this means that clicking on the bu button triggers the click event and thus a call to the named function, clicked().

In the clicked() function, the inputField variable references the input field. The value property of the input field contains the text entered, and this text is displayed in the dialog box.

The outputField variable references the input field that is used for output. The value property can also be assigned a value, and this changes the content of the input field.

Note

You can find the u_form exercise in bonus chapter 1, section 1.11, in the downloadable materials for this book at *www.rheinwerk-computing.com/5875.*

Note

Since the early days of JavaScript, different techniques have been used for the connection between the element of a document and the program code. Modern browsers are familiar with the getElementById() and addEventListener() methods and their many possible uses.

4.2   Submitting and Resetting

With regard to a form, the *submit* and *reset* events can occur. Data can be submitted to a web server using forms, and the resetting process sets the values of forms to their initial values.

The following example responds to these two events. In addition, the content of the form is sent to a file that contains a PHP program, which is located on a web server and processes the data further. In this example, the transferred data is only displayed on the screen.

PHP is a widely used programming language for use on web servers, but you don't need to learn PHP to understand the example. It's only a matter of clarifying the complete process.

4.2.1   The Submit Process

In Figure 4.4, you can see the call of the *form_submit.htm* file. As in the previous examples, it's called from the directory of the JavaScript programs. In my case, that's *C:\js.*

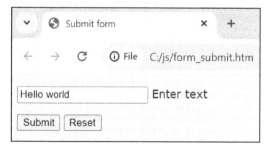

Figure 4.4 Calling Form from Directory

Figure 4.5 shows the response of JavaScript to the submission.

Figure 4.5 Response of JavaScript to Submission

Since the entire process doesn't take place on a web server, the PHP file can't be called properly. Only your code is displayed (see Figure 4.6). A workaround for this problem follows in Section 4.2.2.

```html
<!DOCTYPE html><html lang="en-us">
<head>
    <meta charset="utf-8">
    <title>Response from the web server</title>
    <link rel="stylesheet" href="js5.css">
</head>
<body><p>
<?php
    echo "Entry: " . htmlentities($_POST["entry"]) . "<br>";
?></p>
</body>
</html>
```

Figure 4.6 PHP Code of Response

If you reset the form instead of submitting it, the response will be as shown in Figure 4.7.

Figure 4.7 Response to Resetting Form

4.2.2 A Web Server as an Alternative

If you want to follow the entire transmission process on a web server, there are two alternatives available:

- You can use a local web server on your computer and call the form via the web server. For this purpose, you need to install the freely available XAMPP program package and save the two files involved—*form_submit.htm* and *form_submit.php*—in a directory of the web server. This can be the *jse* directory below the base directory, for example. After calling the *form_submit.htm* file and submitting the form, the response from the PHP program will look like the one shown in Figure 4.8. The installation and use of XAMPP are described in Appendix A, Section A.2.

Figure 4.8 Response via Local Web Server

- Do you have the option of storing data on an internet server? If so, you can upload the two files involved—*form_submit.htm* and *form_submit.php*—to a directory on the internet server. I have saved the files in the *jse* directory on my domain, *theisweb.de*. The address of the directory is *https://theisweb.de/jse*, where you'll find hyperlinks to all the programs in this book, arranged according to their appearance in the chapters. The other examples in this chapter are all accessed via my domain, even if this is not necessary for all programs.

After calling the *form_submit.htm* file and submitting the form, the response of the PHP program looks as shown in Figure 4.9.

Figure 4.9 Response via Internet Server

If you don't use either of these two alternatives and only call the programs with the forms from the directory, it's not a problem. You can use the JavaScript code in any case.

4.2.3 Code for Submitting

Here's the code of the file with the submitted form:

```
... <head> ...
   <script>
     function submit()
     {
        const tx = document.getElementById("idText");
        const entry = tx.value;
        alert("Form will be sent\nEntry: " + entry);
     }

     function reset()
     {
        alert("Form will be reset");
     }
   </script>
</head>
<body>
   <form id="idForm" method="post" action="form_submit.php">
      <p><input id="idText" name="entry"> Enter text</p>
      <p><input type="submit"> <input type="reset"></p>
   </form>
   <script>
      const fo = document.getElementById("idForm");
      fo.addEventListener("submit", submit);
      fo.addEventListener("reset", reset);
   </script>
</body></html>
```

Listing 4.2 form_submit.htm File

The form has the unique ID idForm, and the post value for the method attribute is used to define the secure transmission method *post*, which is selected in all examples. The value of the action attribute references the responding PHP program, which we assume is in the same directory.

The input field has the unique ID idText, which is required to access the content of the field using JavaScript. The input field also has the name attribute, here with the entry value, which is required to access the content of the field using PHP.

The submit and reset values for the type attribute of the two input elements identify them as buttons for submitting or resetting the form.

The fo variable references the form, and the addEventListener() method is used to link the submit and reset events with the named functions submit() and reset().

4.2.4 Code for Receiving

As already mentioned, you don't need to learn PHP—but for information and clarification purposes, the code of the receiving PHP file as you have already seen it in Figure 4.6 follows here *as an exception*:

```
<!DOCTYPE html><html lang="en-us">
<head>
   <meta charset="utf-8">
   <title>Response from the web server</title>
   <link rel="stylesheet" href="js5.css">
</head>
<body><p>
<?php
   echo "Entry: " . htmlentities($_POST["entry"]) . "<br>";
?></p>
</body>
</html>
```

Listing 4.3 form_submit.php File

This is a standard HTML document, and the PHP code is embedded in a container between `<?php` and `?>`. The language element `echo` is used for output on the screen, and the dots connect the individual text parts with each other, like the plus sign (+) in JavaScript.

Those elements that have the `name` attribute in the form are automatically turned into elements of the PHP field `$_POST` with their attribute values. Here's what it looks like: `name="entry"` becomes `$_POST["entry"]`.

The `htmlentities()` function is used to protect against the effect of malicious code that a malicious user could transmit to the web server using the input field.

There's a responding PHP file for almost all the form examples in this book to illustrate the process.

4.3 Mandatory Fields and Checking

You can easily validate the entries on a form (i.e., check them for validity) using HTML. Further validation of the entered or selected contents of a form is possible using JavaScript.

In the following example, a family name and a password must be entered in two mandatory fields (see Figure 4.10), and a comment can be entered in a multiline input field. The following happens when you click the **Submit** button:

- If one of the two mandatory fields is empty, an error message is displayed within the form (see Figure 4.11) and the form is not submitted.

- If a password with less than three characters or more than eight characters is entered, an error message is displayed by JavaScript (see Figure 4.12) and the form is not submitted.

- If both mandatory fields are filled in and the password has the correct length, the form is submitted.

As soon as you have more knowledge of the structure and handling of strings, you can carry out further checks. In Chapter 6, Section 6.2.5, you can find a longer example.

Figure 4.10 Mandatory Fields and Multiline Input Field

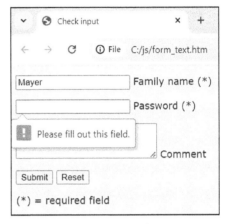

Figure 4.11 Error Message from HTML

Figure 4.12 Error Message from JavaScript

Here's the code of the file that contains the form:

```
... <head> ...
  <script>
    function submit(e)
    {
        const pw = document.getElementById("idPassword").value;
```

```
        if(pw.length < 3 || pw.length > 8)
        {
            alert("Password must contain 3-8 characters");
            e.preventDefault();
            return false;
        }

        alert(document.getElementById("idFamilyName").value
            + "\n" + document.getElementById("idPassword").value
            + "\n" + document.getElementById("idComment").value);
    }
  </script>
</head>
<body>
  <form id="idForm" method="post" action="form_text.php">
    <p><input id="idFamilyName" name="familyName"
        required="required"> Family name (*)</p>
    <p><input id="idPassword" name="password"
        type="password" required="required"> Password (*)</p>
    <p><textarea id="idComment" rows="3" cols="25"
        name="comment">(Contents)</textarea> Comment</p>
    <p><input type="submit"> <input type="reset"></p>
    <p>(*) = required field</p>
  </form>
  <script>
    document.getElementById("idForm").addEventListener
        ("submit", function(e) { return submit(e);});
  </script>
</body></html>
```

Listing 4.4 form_text.htm File

The structure of the form looks as follows:

The password value for the type attribute of an input element turns this element into an input field for a password. Only the number of characters entered can be recognized—the characters themselves can't be read.

A multiline input field is displayed using a textarea container. The size of the field initially depends on the values of the rows and cols attributes, but it can usually be changed in the browser. The entry is available in JavaScript via the value property, as with single-line input fields.

The two input fields for the family name and password have the required attribute with the required value, so the fields must not be left blank. If neither of the two input fields is empty, the submit event gets triggered.

The process in the event handler is as follows:

1. The addEventListener() method is called directly for the reference to the form, and this abbreviated notation is also used in most of the following examples.

2. The transfer of a parameter when linking an event to a function is only possible using an anonymous function. In this example, the parameter is a reference to the event object, which is automatically available with more information when an event occurs. According to convention, the reference is given the name e.

3. In the anonymous function, the named submit() function is called. The parameter of the anonymous function is passed in turn, and the submit() function returns a truth value to the anonymous function. If this is false, the form won't be submitted.

The process in the submit() function is as follows:

1. In the submit() function, the password entered is determined by calling the value property directly for the reference to the input field. This is also an abbreviated notation, which is also used in the following examples.

2. The length property of a string contains the number of characters in a string. You can find out more about strings in Chapter 6, Section 6.2. If the password is too short or too long, the submit() function returns the value false and an error message is displayed.

3. In addition, the preventDefault() method of the event object is called so that the default action for the element (in this case, submitting the form) is not carried out.

4. If the password has the correct length, the three entries are displayed, the submit() function returns the true value, and the form is submitted.

4.4   Radio Buttons and Checkboxes

A group of radio buttons lets users choose different options, and a checkbox lets users make an additional selection. You can use JavaScript to determine which selection has been made.

In the following example, one of three different countries can be selected and two other countries can also be selected (see Figure 4.13).

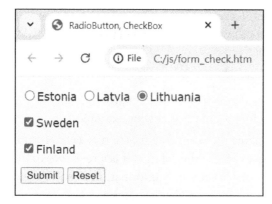

Figure 4.13 Radio Buttons and Checkboxes

When the form is submitted, the values of the various selection elements are displayed (see Figure 4.14). They don't have to match the displayed texts.

Figure 4.14 Values of Selection Elements

Here's the code of the file that contains the form:

```
... <head> ...
  <script>
    function submit()
    {
      let tx;

      if(r1.checked) tx = r1.value;
      else if(r2.checked) tx = r2.value;
      else if(r3.checked) tx = r3.value;

      if(c1.checked) tx += "\n" + c1.value;
      if(c2.checked) tx += "\n" + c2.value;

      alert(tx);
    }
  </script>
```

```
</head>
<body>
   <form id="idForm" method="post" action="form_check.php">
      <p><input id="idEstonia" name="balticCountries" type="radio"
            value="Estonia" checked="checked">Estonia
         <input id="idLatvia" name="balticCountries" type="radio"
            value="Latvia">Latvia
         <input id="idLithuania" name="balticCountries" type="radio"
            value="Lithuania">Lithuania</p>
      <p><input id="idSweden" name="sweden" type="checkbox"
            value="Sweden">Sweden</p>
      <p><input id="idFinland" name="finland" type="checkbox"
            value="Finland" checked="checked">Finland</p>
      <p><input type="submit"> <input type="reset"></p>
   </form>
   <script>
      const r1 = document.getElementById("idEstonia");
      const r2 = document.getElementById("idLatvia");
      const r3 = document.getElementById("idLithuania");
      const c1 = document.getElementById("idSweden");
      const c2 = document.getElementById("idFinland");
      document.getElementById("idForm")
         .addEventListener("submit", submit);
   </script>
</body></html>
```

Listing 4.5 form_check.htm File

The structure of the form looks like this:

- The radio and checkbox attributes identify the input elements as radio buttons and checkboxes, respectively. The value attribute contains the value that is transmitted for the element.

- All radio buttons with the same value for the name attribute belong together. If a selection is made within the group, the previous selection will be canceled.

- One of the radio buttons within a group should already be marked using the checked attribute and the checked value so that the group always has a value.

The references to the individual elements are saved.

The process in the submit() function is as follows: The group of radio buttons is checked using multiple branches, and the checkboxes are each checked using a single branch. If the checked property has the value true, the relevant element gets selected. The value property also contains the value of the element here.

4.5   Selection Menus

You can use a single-selection menu to choose among different options, and you can select several options simultaneously in a multiple-selection menu using the ⬆ and Ctrl keys. You can also use JavaScript to determine which selection has been made.

The following example shows a single-selection menu and a multiple-selection menu. Each menu contains three countries (see Figure 4.15).

Figure 4.15 Single- and Multiple-Selection Menus

When the form gets submitted, the values of the various selection elements are displayed (see Figure 4.16). These don't have to match the displayed texts.

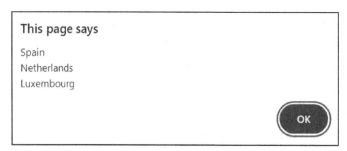

Figure 4.16 Values of Selection Elements

Here's the code of the file that contains the form:

```
...  <head> ...
   <script>
      function submit()
      {
         let tx = document.getElementById("idSingle").value;

         const opt = document.getElementById("idMultiple").options;
         for(let i=0; i<opt.length; i++)
            if(opt[i].selected)
               tx += "\n" + opt[i].value;

         alert(tx);
      }
   </script>
</head>
<body>
   <form id="idForm" method="post" action="form_select.php">
      <p>Country 1: <select id="idSingle" name="single">
            <option value="Italy" selected="selected">Italy</option>
            <option value="Spain">Spain</option>
            <option value="Romania">Romania</option>
         </select></p>
      <p>Countries 2 to 4:
         <select id="idMultiple" name="multiple[]" multiple="multiple">
            <option value="Belgium" selected="selected">Belgium</option>
            <option value="Netherlands">Netherlands</option>
            <option value="Luxembourg" selected="selected">Luxembourg</option>
         </select></p>
      <p><input type="submit"> <input type="reset"></p>
   </form>
   <script>
      document.getElementById("idForm").addEventListener("submit", submit);
   </script>
</body></html>
```

Listing 4.6 form_select.htm File

The structure of the form looks like this:

- A selection menu is available in a select container, and the multiple attribute and the multiple value turn it into a multiple-selection menu. The individual options are available in option containers, and the values of the individual options are defined using the value attribute.

- As with the radio buttons, one of the options in a simple selection menu should be selected using the selected attribute and the selected value so that the menu always has a value.

In the case of a multiple-selection menu, it's advisable to add square brackets to the value for the name attribute. This means that the individual elements can be treated like field elements when they are parsed in the PHP program.

The process in the submit() function is as follows:

1. The value of a simple selection menu is easy to determine as only one option can be selected. The value property of the menu contains the desired value.

2. With the multiple-selection menu, the determination is more complex, as multiple values can be selected. The options property of a menu is a field with references to the individual options. The length property corresponds to the size of the field (i.e., the number of options), and a single option can be accessed using square brackets. Each option has the selected property. If the option is selected, the property has the value true; otherwise, it has the value false. The value property of an option contains its value.

> **Note**
>
> You can find the u_select exercise in bonus chapter 1, section 1.13, in the downloadable materials for this book at *www.rheinwerk-computing.com/5875*.

4.6 Other Form Events

You have already become familiar with the click (on a button) and submit and reset (of a form) events. The following program adds the following events that can take place within a form:

- click: Clicking on a radio button or a checkbox
- change: Changing the content of an input field or the selection in a selection menu
- keyup: Releasing a key in an input field
- focus: Accessing an element (i.e., it receives the input focus, and the input cursor is positioned in the element)
- blur: Exiting an element (i.e., it loses the input focus, and the input cursor is no longer positioned in the element)

The responses to the various events are displayed in a read-only input field. As an example, in Figure 4.17, you can see the output after clicking on a checkbox to select it.

Figure 4.17 Response to Marking Checkbox

Here's the code of the file that contains the form:

```
... <head> ...
   <script>
      function value(e)
      {
         const re = document.getElementById("idResponse");
         re.value = e.type + ", ";

         const id = e.target.id;
         if(id == "idTrip")
            if(document.getElementById(id).checked)
               re.value += "Marker set";
            else
               re.value += "Marker deleted";
         else
            re.value += document.getElementById(id).value;
      }
   </script>
</head>
<body id="idBody">
   <form method="post" action="form_event.php">
```

```
      <p><input id="idMr" name="address" type="radio"
          value="Mr" checked="checked"> Mr
        <input id="idMs" name="address" type="radio"
          value="Ms"> Ms</p>
      <p><input id="idFamilyName" name="familyName"> Family name</p>
      <p><input id="idFirstName" name="firstName"> First name</p>
      <p><input id="idCity" name="city"> City</p>
      <p><select id="idCountry" name="country">
          <option value="Italy" selected="selected">Italy</option>
          <option value="Spain">Spain</option>
          <option value="Portugal">Portugal</option>
        </select> Country</p>
      <p><input id="idTrip" name="trip" type="checkbox"
        value="OneWayOnly"> one-way only</p>
      <p><input type="submit"> <input type="reset"></p>
      <p><input id="idResponse"
        readonly="readonly"> Response</p></p>
    </form>
    <script>
      document.getElementById("idMr").addEventListener
        ("click",  function(e) { value(e); });
      document.getElementById("idMs").addEventListener
        ("click",  function(e) { value(e); });
      document.getElementById("idFamilyName").addEventListener
        ("change", function(e) { value(e); });
      document.getElementById("idFirstName").addEventListener
        ("keyup",  function(e) { value(e); });
      document.getElementById("idCity").addEventListener
        ("focus",  function(e) { value(e); });
      document.getElementById("idCity").addEventListener
        ("blur",  function(e) { value(e); });
      document.getElementById("idCountry").addEventListener
        ("change", function(e) { value(e); });
      document.getElementById("idTrip").addEventListener
        ("click",  function(e) { value(e); });
    </script>
</body></html>
```

Listing 4.7 form_event.htm File

The structure of the form and the event handler are as follows:

- The readonly attribute with the readonly value ensures that no entry is possible in the input field. It's only used to output information.

- The click event is triggered when one of the two radio buttons is selected. The event is linked to an anonymous function that calls the named function, value(), and the e parameter, which refers to an event object, is passed on. Other events are linked to the value() function in the same way.

- In an input field, the change event takes place after the content of the input field has changed and as soon as another element is selected or clicked outside the input field.

- The keyup event in an input field is triggered immediately after a key has been released. This event occurs after a character has been entered or deleted.

- In a selection menu, the change event takes place as soon as the selection changes. If a checkbox is checked or unchecked, the click event gets triggered.

- The focus and blur events take place in an input field as soon as the input field is accessed or subsequently exited.

The process in the value() function is as follows:

1. The target property of the event object references the element for which the event was triggered. The type property contains the name of the event, and the id sub-property of the target property corresponds to the unique ID of the element. In this way, you can use one function for multiple elements or events.

2. The checked property of the checkbox is checked to determine whether the check-mark was set or removed at the click event. For all other elements, the current value of the element for which the event was triggered gets displayed.

You should test the possible events once and note the time at which the information appears and the content of the response.

4.7 Mouse Events

In addition to forms, there are events that can be triggered using the mouse. The following program handles the following events:

- click: Clicking on an element
- mousemove: Moving the mouse over an element
- mousedown: Pressing down a mouse button over an element
- mouseup: Releasing a mouse button over an element
- mouseover: Accessing an element
- mouseout: Exiting an element

To illustrate this, these events take place in images. Information on the events is displayed in a read-only input field (see Figure 4.18).

Figure 4.18 Mouse Events

Here's the code of the file:

```
... <head> ...
  <script>
    function mouse(e)
    {
      document.getElementById("idResponse").value = e.type + " " + e.offsetX
        + "/" + e.offsetY + " " + document.getElementById(e.target.id).src;
    }
  </script>
</head>
<body>
  <form>
    <p><input size="70" id="idResponse" readonly="readonly"></p>
  </form>
  <p><img src="im_paradise.jpg"       id="idParadise"     alt="Paradise">
    <img src="im_solar_eclipse.jpg" id="idSolarEclipse" alt="Solar eclipse">
    <img src="im_wave.jpg"          id="idWave"         alt="Wave"></p>
  <p><img src="im_winter.jpg"       id="idWinter"       alt="Winter">
    <img src="im_tree.jpg"          id="idTree"         alt="Tree"></p>

  <script>
    document.getElementById("idParadise").addEventListener
```

```
        ("click", function(e) { mouse(e); });
    document.getElementById("idSolarEclipse").addEventListener
        ("mousemove", function(e) { mouse(e); });
    document.getElementById("idWave").addEventListener
        ("mousedown", function(e) { mouse(e); });
    document.getElementById("idWave").addEventListener
        ("mouseup", function(e) { mouse(e); });

    document.getElementById("idWinter").addEventListener
        ("mouseover", function(e) { mouse(e); });
    document.getElementById("idWinter").addEventListener
        ("mouseout", function(e) { mouse(e); });

    const tree = document.getElementById("idTree");
    tree.addEventListener("mouseover",
        function() { tree.src = "im_paradise.jpg"; });
    tree.addEventListener("mouseout",
        function() { tree.src = "im_tree.jpg"; });
  </script>
</body></html>
```

Listing 4.8 mouse_event.htm File

The structure of the event handler looks like this:

- Clicking on the first image triggers the click event, which is connected to an anonymous function that calls the named function, mouse(). The e parameter, which refers to an event object, is passed on, and other events are also linked to the mouse() function in the same way.

- An img element has the src property, which contains the name of the image file, including the path if applicable. In the last image, an image change is performed by calling an anonymous function for the mouseover or mouseout events. The src property is given a new value within the function.

Three pieces of information are output in the mouse() function:

- The name of the mouse event
- The relative coordinates (offsetX and offsetY) of the location within the element at which the mouse event took place
- The name of the image file that is displayed in the relevant img element

You should test the possible events once and note the location of the event, the associated information, and the time at which the information appears.

> **Note**
>
> You can find the u_mouse exercise in bonus chapter 1, section 1.14, in the downloadable materials for this book at *www.rheinwerk-computing.com/5875.*

4.8   Changing the Document

By default, hyperlinks are used to switch to another document in the browser. Using the predefined location object and the appropriate events, this can also be done differently.

In the following program (see Figure 4.19) this is done in three ways:

- After clicking on a button
- After clicking on an image
- By changing the selection in a selection menu

The predefined location object contains information about the current URL, such as an HTML document.

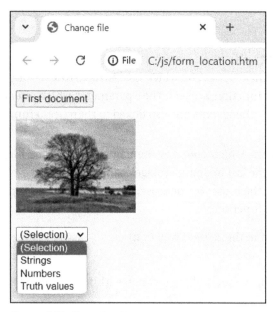

Figure 4.19 Changing Document

Here's the code of the file:

```
...
<body>
    <form>
```

```
    <p><input id="idButton" type="button" value="First document"></p>
    <p><img src="im_tree.jpg" id="idTree" alt="Tree"></p>
    <p><select id="idSelection">
        <option value="(Selection)" selected="selected">(Selection)</option>
        <option value="strings.htm">Strings</option>
        <option value="numbers.htm">Numbers</option>
        <option value="bool.htm">Truth values</option>
    </select> </p>
</form>
<script>
   document.getElementById("idButton").addEventListener
       ("click", function() { location.href = "first.htm"; } );
   document.getElementById("idTree").addEventListener("click",
       function() { location.href = "special_characters.htm"; } );

   const selection = document.getElementById("idSelection");
   selection.addEventListener("change", function() {
       if(selection.value != "(Selection)")
           location.href = selection.value; } );
</script>
</body></html>
```

Listing 4.9 form_location.htm File

The click event on the button is linked to an anonymous function that assigns a file name as a new value to the href property of the location object. This causes the switch to the desired document, and the same thing happens when you click on the image.

The change event is also linked to an anonymous function when you change the selection in the selection menu. This function checks whether the value of the currently selected option (i.e., the value of the entire selection menu) corresponds to the value of the first option. The first option serves as the title of the selection menu and should not lead to a document change. If the selected option is not the first option, the system switches to the desired document.

4.9   Other Types and Properties

There are other types and properties of input fields that have been made available, especially since version 5 of HTML. They provide a wide range of easy-to-use options.

The various types can contain a validation (i.e., a check of the element when the form is submitted). This serves to avoid the transmission of incorrect data, and this capability can supplement or even replace form checks with JavaScript. Before creating a Java-Script program, you should check whether the desired validation can't already be carried out using HTML in modern browsers.

If an individual browser doesn't implement a certain type of input field, a standard input field will be displayed instead. This allows an entry to be made in every case. Mobile browsers in particular use numerous features and support you by displaying a keyboard with suitable elements.

4.9.1 Text Inputs, Search Fields, and Colors

The following program presents input fields for various text types, search fields with lists, and elements for selecting colors. The use of special properties is also explained.

You can see the entire document in Figure 4.20.

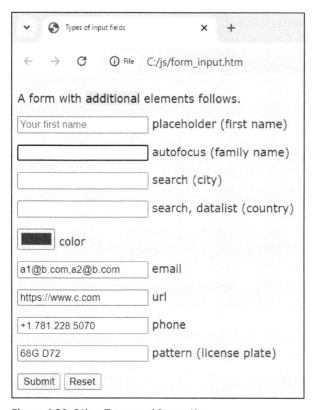

Figure 4.20 Other Types and Properties

Here's the code of the file:

```
... <head> ...
  <script>
    function submit()
```

```
      {
         alert(document.getElementById("idFirstName").value
            + "\n" + document.getElementById("idFamilyName").value
            + "\n" + document.getElementById("idZip").value
            + " " + document.getElementById("idCity").value
            + "\n" + document.getElementById("idCountry").value
            + "\n" + document.getElementById("idColor").value
            + "\n" + document.getElementById("idAddresses").value
            + "\n" + document.getElementById("idWeb").value
            + "\n" + document.getElementById("idPhone").value
            + "\n" + document.getElementById("idLicensePlate").value);
      }
   </script>
</head>
<body>
   <p>A form with <mark>additional</mark> elements follows.</p>
   <form id="idForm" method="post" action="form_input.php">
      <p><input id="idFirstName" name="firstName"
         placeholder="Your first name"> placeholder (first name)</p>
      <p><input id="idFamilyName" name="familyName" autofocus>
         autofocus (family name)</p>
      <input id="idZip" name="zip" type="hidden" value="02171">

      <p><input id="idCity" name="city" type="search"> search (city)</p>
      <p><input id="idCountry" name="country" type="search"
         list="idList"> search, datalist (country)</p>
      <datalist id="idList">
         <option value="Germany">
         <option value="France">
         <option value="Italy">
      </datalist>

      <p><input id="idColor" name="color" type="color" value="#ff0000"> color</p>
      <p><input id="idAddresses" name="addresses" type="email"
         multiple="multiple" value="a1@b.com, a2@b.com"> email</p>
      <p><input id="idWeb" name="web" type="url" value="https://www.c.com"> url</p>
      <p><input id="idPhone" name="phone" type="tel"
         value="+1.781.228.5070"> phone</p>
      <p><input id="idLicensePlate" name="licensePlate" value="68G D72"
         pattern="^[1-9][0-9][A-Z] [A-Z][0-9]{1,2}$">
         pattern (license plate)</p>
      <p><input type="submit"> <input type="reset"></p>
   </form>
```

```
<script>
    document.getElementById("idForm").addEventListener("submit", submit);
</script>
</body></html>
```

Listing 4.10 form_input.htm File

A mark container is used to visually highlight text (e.g., with yellow highlighting). It can be used to mark the important elements of a form.

In the element for the first name, a placeholder appears in a light gray color using the placeholder attribute. A placeholder is used to explain an input field, and if an entry is started in the input field, the placeholder disappears.

After calling the program, the input cursor automatically appears in the element for the family name, thanks to the autofocus attribute. You can immediately start entering data there, and if there's already text in such an element, it will be highlighted in full.

It has long been possible to hide an input field using the hidden value for the type attribute. In this way, additional data that should not be visible during use can be transmitted to the web server.

The search value for the type attribute identifies an input field that can be recognized as a typical search field. It can also contain an icon for deleting the search term.

A search field can have the additional list attribute, whose value corresponds to the ID of a list with predefined entries that can be selected when operating the search field (see also Figure 4.21). As soon as characters are entered in the search field, only those entries appear in which these characters occur. Additional search terms can also be entered. The list is created using a datalist container, and each entry represents the value of the value attribute of an option tag.

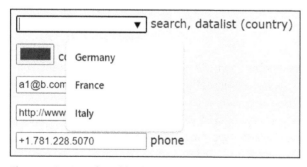

Figure 4.21 Search Field with List

The color value for the type attribute indicates an input field for setting a color. Once the input field has been selected, a dialog box with setting options opens (see also Figure 4.22).

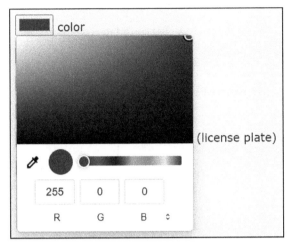

Figure 4.22 Setting Color

An email address can be entered in an input field that has the email value for the type attribute. The field can be empty, and if it's not empty, it must contain the minimum components of an email address or the form won't be submitted. If the multiple attribute with the multiple value gets added, then multiple addresses can be entered, separated by commas.

The situation is similar for an input field with the url value for the type attribute. The field can be empty, and if it's not empty, it must contain the minimum components of a URL or the form won't be submitted. However, multiple URLs can't be entered.

The element for entering a phone number has the tel value for the type attribute. It can provide the option of accessing existing phone lists.

The pattern attribute is used to validate an input using regular expressions. If an entry is made that doesn't match the pattern, the form can't be submitted.

In this case, we have an example with a license plate. We assume that it starts with two digits, the first of which must not be 0. This is followed by a capital letter, a space and another capital letter. This is followed by 1 or 2 digits at the end of the license plate. You can find out more about regular expressions in bonus chapter 2 in the downloadable materials for this book at *www.rheinwerk-computing.com/5875*.

After submitting, the submit() function is called. All values are displayed using the value property.

4.9.2   Elements for Numbers

This section describes some elements that can be used for secure input and the clear representation of numerical values (see Figure 4.23).

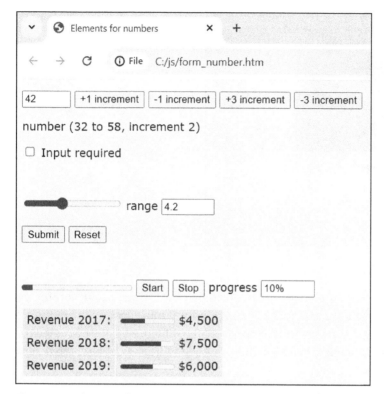

Figure 4.23 Elements for Numbers

Here's the content of the form in the middle part of the file:

```
...
<body>
   <form id="idForm" method="post" action="form_number.php">
      <p><input id="idNumber" name="number" type="number"
         min="32" max="58" step="2" value="42">
        <input id="idPlus1" type="button" value="+1 increment">
        <input id="idMinus1" type="button" value="-1 increment">
        <input id="idPlus3" type="button" value="+3 increment">
        <input id="idMinus3" type="button" value="-3 increment"></p>
      <p>number (32 to 58, increment 2)</p>
      <p><input id="idRequired" type="checkbox"> Input required</p><br>
      <p><input id="idRange" name="range" type="range" min="3.2"
         max="5.8" step="0.2" value="4.2"> range <input id="idRValue"
         readonly="readonly" size="5" value="4.2"></p>
      <p><input type="submit"> <input type="reset"></p><br>

      <p><progress id="idProgress" max="100" value="10"></progress>
        <input id="idStart" type="button" value="Start">
```

```
      <input id="idStop" type="button" value="Stop"> progress <input
          id="idPValue" readonly="readonly" size="5" value="10%"></p>
  </form>
...
```

Listing 4.11 form_number.htm File, Middle Part

The element for entering a numerical value has the number value for the type attribute. If you place the cursor in the input field, an additional up/down button is displayed, and you can change the number by clicking on it. To set the element, you can enter values for the min, max, step, and value attributes. These indicate the lower limit of 32, the upper limit of 58, the increment of 2, and the displayed value of 42. A value that is not within the limits or doesn't match the increment won't be submitted, and in that case, an error message will display (see Figure 4.24).

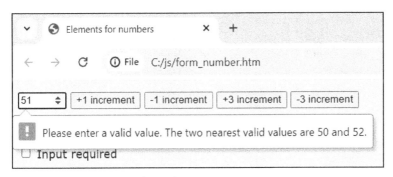

Figure 4.24 Not Permitted Number

You can also set the value via code. The **+1 increment**, **−1 increment**, **+3 increment**, and **−3 increment** buttons change the value by the corresponding multiple of the increment of 2 (e.g., the **+3 increment** button changes the value from 42 to 42 + 3 × 2 = 48). The required attribute can also be changed via code. If you select the checkbox, the number field can no longer be left blank.

A *slider* is used to securely transmit a numerical value that originates from a range. This element has the range value for the type attribute. The value itself is not displayed, so a separate display is added here in a read-only input field. The min, max, step, and value attributes are also available for setting purposes, and numbers with decimal places can also be set using the slider.

The progress container is usually used to display a percentage progress. The min and max attributes have the default values 0 and 1. Here, 10 is taken as the current value and 100 as the value for the maximum. The **Start** and **Stop** buttons are used to start a JavaScript function that animates the progress bar, and the current value is displayed in a read-only input field.

Let's take a look at the lower part of the document:

```
...
   <table>
      <tr><td>Revenue 2017:</td><td><meter min="0" max="10000"
            value="4500"></meter> $4,500</td></tr>
      <tr><td>Revenue 2018:</td><td><meter min="0" max="10000"
            value="7500"></meter> $7,500</td></tr>
      <tr><td>Revenue 2019:</td><td><meter min="0" max="10000"
            value="6000"></meter> $6,000</td></tr>
   </table>

   <script>
      document.getElementById("idForm").addEventListener("submit", submit);

      const number = document.getElementById("idNumber");
      document.getElementById("idRequired").addEventListener
         ("click", function() { number.required = !number.required; } );
      document.getElementById("idPlus1").addEventListener
         ("click", function() { number.stepUp(); } );
      document.getElementById("idMinus1").addEventListener
         ("click", function() { number.stepDown(); } );
      document.getElementById("idPlus3").addEventListener
         ("click", function() { number.stepUp(3); } );
      document.getElementById("idMinus3").addEventListener
         ("click", function() { number.stepDown(3); } );

      document.getElementById("idRange").addEventListener
         ("change", changeRange);
      document.getElementById("idStart").addEventListener
         ("click", startProgress);
      document.getElementById("idStop").addEventListener
         ("click", stopProgress);
   </script>
</body></html>
```

Listing 4.12 form_number.htm File, Lower Part

A meter container is actually used to display a filling level. Here, three meter containers are used to display revenues using the min, max, and value attributes, similar to in a bar chart.

The number variable is used to reference the number input field.

After activating the checkbox, the value of the Boolean property required is reversed using the logical negation (!) operator. If the checkbox is selected, the number input field becomes a mandatory field and must not remain empty.

The stepUp() and stepDown() methods change the value of the element by the corresponding multiple of the increment. In the absence of a parameter, the value will be changed by exactly one increment.

Moving the slider leads to the change event and thus to the call of the changeRange() function. This ensures that the current value will be displayed next to the slider.

The **Start** button starts the startProgress() function, which changes the progress bar. Pressing the **Stop** button calls the stopProgress() function, in which the change is ended. This ensures that the current value will be displayed next to the progress bar.

Now, here's the JavaScript code from the first part of the document:

```
...  <head> ...
  <script>
     function submit()
     {
        alert(number.value + "\n"
           + document.getElementById("idRange").value);
     }

     function changeRange()
     {
        const range = parseFloat(
           document.getElementById("idRange").value);
        document.getElementById("idRValue").value
           = range.toFixed(1);
     }

     let pValue = 10;
     let reference;

     function startProgress()
     {
        if(pValue<100)
        {
           pValue++;
           document.getElementById("idProgress").value = pValue;
           document.getElementById("idPValue").value = pValue + "%";
              reference = setTimeout(startProgress, 20);
        }
     }
```

```
      function stopProgress()
      {
         clearTimeout(reference);
      }
   </script>
</head>
...
```

Listing 4.13 form_number.htm File, JavaScript Code

The value of the number input field is output in the submit() function, and the change-Range() function is used to output the numerical value of the slider with one decimal place.

The setTimeout() method of the window object results in a timeout. A function is started with a time delay, and a reference to the relevant function (here, startProgress()) and the desired time delay in milliseconds are used as parameters. A reference is returned, and it's saved in the reference variable, which is valid throughout the entire program. As the startProgress() function calls itself again and again, the progress bar is animated in small steps from the start value 10 to the end value 100.

An already scheduled next call of the startProgress() function can be prevented by calling the clearTimeout() function. This requires the reference returned by the set-Timeout() function as a parameter, and this terminates the animation prematurely. You can find out more about time-controlled processes in Chapter 6, Section 6.5.

The pValue variable, which is valid throughout the entire program, contains the current value of the progress bar.

4.9.3   Elements for Time Information

This section describes some elements that can be used for the secure setting and transmission of time data (see Figure 4.25). After you select one of the elements, as shown in Figure 4.26, control elements appear for a convenient entry or selection of the individual parts of the time specification within a defined time range, as shown in Figure 4.27.

Let's start with the code of the form:

```
...
<body>
   <form id="idForm" method="post" action="form_time.php">
      <p>1: <input id="idDate" name="date" type="date" value="2024-05-25"
         min="2024-04-15" max="2024-06-20"> date</p>
      <p>2: <input id="idTime" name="time" type="time"
         value="10:30" min="08:00" max="13:00"> time</p>
```

```
    <p>3: <input id="idLocale" name="locale" type="datetime-locale"
        value="2024-05-25T10:30"> datetime-locale</p>
    <p>4: <input id="idMonth" name="month" type="month" value="2024-05"
        min="2023-10" max="2025-03"> month</p>
    <p>5: <input id="idWeek" name="week" type="week" value="2024-W21"
        min="2024-W10" max="2024-W30"> week</p>
    <p><input type="submit"> <input type="reset"></p>
  </form>
  <script>
    document.getElementById("idForm").addEventListener("submit", submit);
  </script>
</body></html>
```

Listing 4.14 form_time.htm File, Form with Elements

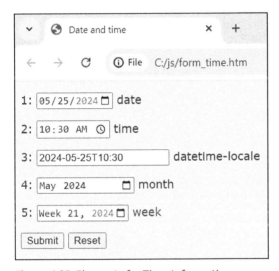

Figure 4.25 Elements for Time Information

All elements have the `input` type with different values for the `type` attribute: `date`, `time`, `datetime-locale`, `month`, and `week`. The `min` and `max` attributes are used to limit the time range, and the `value` attribute stands for the default setting of the time specification. For the five different elements, the appropriate output format is used for this default setting.

The first `input` element of the `date` type is used to set a date with a clock time of 00:00 *Coordinated Universal Time* (UTC), which is coordinated world time. You can select individual parts of the time entry for input (see Figure 4.26).

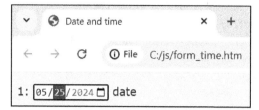

Figure 4.26 Selection of Part of Time Specification

If no date is preset, the current month appears—but in this case, no date gets transmitted when the form is submitted, so a presetting is recommended. Data that is outside the permitted range is displayed in light gray and can't be selected (see Figure 4.27).

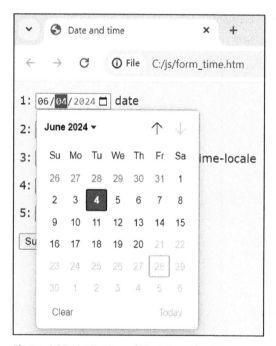

Figure 4.27 Limitation of Time Specification

The second `input` element of the `time` type is used to set a clock time with the date of 01/01/1970.

The element of the `datetime-locale` type can be used to set both the date and the clock time. Note the separate format for the `value` attribute.

The `month` type is selected for the fourth element, in which only entire months can be set. The fifth element has the `week` type, in which only entire calendar weeks can be set.

Here's the structure of the submit() function:

```
...  <head> ...
    <script>
      function submit()
      {
          const d = document.getElementById("idDate");
          const dValue = d.value;
          const dDate = d.valueAsDate;
          const dLocale = dDate.toLocaleDateString();

          const t = document.getElementById("idTime");
          const tValue = t.value;
          let tDate = t.valueAsDate;
          tDate.setHours(tDate.getHours() - 1);
          const tLocale = tDate.toLocaleTimeString();

          const cValue = document.getElementById("idLocale").value;
          const mValue = document.getElementById("idMonth").value;
          const wValue = document.getElementById("idWeek").value;

          alert(dValue + " # " + dLocale + "\n" + tValue + " # " + tLocale
              + "\n" + cValue + "\n" + mValue + "\n" + wValue);
      }
    </script>
</head>
...
```

Listing 4.15 form_time.htm File, JavaScript Code

The value of an input field is determined via the value property.

A value for the valueAsDate property can also be determined for the first two input fields. This property contains the value as a date object for which the toLocaleDate-String() and toLocaleTimeString() functions can be called. These functions generate a string with a local time specification.

One hour must be subtracted from the time so that the displayed result matches the set value. The getHours() method returns the value of the hour of a Date object, while the setHours() method changes this value.

You can find out more about the Date object in Chapter 6, Section 6.4.

In Figure 4.28, you can see the display after the form has been submitted.

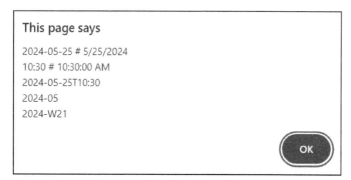

Figure 4.28 Output of Time Data

4.9.4   Validating Forms

JavaScript provides even more options with regard to the validation of form content. You can set whether an individual element or an entire form should be checked or not. There are two forms in the document, each with a **Submit** button (see Figure 4.29).

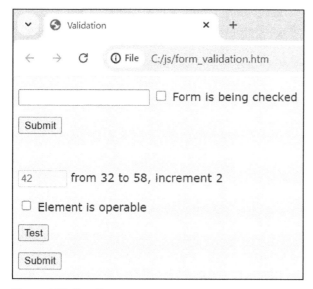

Figure 4.29 Two Forms

The first form contains an input field for text, and you can use a checkbox to determine whether or not you want the form to be checked (e.g., depending on the content entered).

In the second form, you can use a checkbox to determine whether the number input field contained therein should be operable or not. If it can't be operated, the input won't be checked.

In addition, you can use the **Test** button to determine whether the input should be checked and whether it would pass a test. This can be useful if the content of a form is set automatically, and you want to know whether the current content would stand up to a check.

Here's the lower part of the document with the two forms and the event handlers:

```
...
<body>
   <form id="idForm1" novalidate method="post" action="form_validation1.php">
      <p><input id="idContents1" name="contents1" required>
      <input id="idCheck1" type="checkbox"> Form is being checked</p>
      <p><input type="submit"></p>
   </form>
   <p> </p>

   <form id="idForm2" method="post" action="form_validation2.php">
      <p><input id="idContents2" name="contents2" disabled type="number" min="32"
         max="58" step="2" value="42"> from 32 to 58, increment 2</p>
      <p><input id="idCheck2" type="checkbox"> Element is operable</p>
      <p><input id="idTest2" type="button" value="Test"></p>
      <p><input type="submit"></p>
   </form>

   <script>
      document.getElementById("idForm1").addEventListener("submit", submit1);
      document.getElementById("idCheck1").addEventListener("click", check1);

      document.getElementById("idTest2").addEventListener("click", test2);
      document.getElementById("idForm2").addEventListener
         ("submit", function(e) { return submit2(e);});
      document.getElementById("idCheck2").addEventListener("click", check2);
   </script>
</body></html>
```

Listing 4.16 form_validation.htm File, Lower Part

The novalidate attribute is used to determine that the first form should not be checked. In this case, no error message appears, even if the input field (which is actually a mandatory field) is left blank. You can use the checkbox to determine the form to be checked.

The number input field of the second form is initially not operable due to the disabled attribute. The input would pass the test with a value of 42, but only the even numbers from 32 to 58 are valid values.

Now, here's the upper part of the document:

```
... <head> ...
   <script>
      function submit1()
      {
         alert(document.getElementById("idContents1").value);
      }

      function check1()
      {
         const f = document.getElementById("idForm1");
         f.noValidate = !f.noValidate;
      }

      function test2()
      {
         const number = document.getElementById("idContents2");
         alert("Would be tested: " + number.willValidate + "\n"
            + "Too large: " + number.validity.rangeOverflow + "\n"
            + "Too small: " + number.validity.rangeUnderflow + "\n"
            + "Would be valid: " + number.validity.valid);
      }

      function submit2(e)
      {
         const contents2 = document.getElementById("idContents2");
         const number = contents2.valueAsNumber;
         if(contents2.disabled)
         {
            alert("Entry is disabled");
            e.preventDefault();
            return false;
         }
         else
            alert(number);
      }

      function check2()
      {
         document.getElementById("idContents2").disabled =
            !document.getElementById("idContents2").disabled;
      }
```

```
    </script>
</head>
```
. . .

Listing 4.17 form_validation.htm File, JavaScript Code

The check1() function is executed each time the checkbox of the first form is clicked. This sets the value of the noValidate property of the form to true or false. When you send the form, an error message only appears as in Figure 4.30 if noValidate has the false value and the input field is empty. You can therefore specify in your programs whether a form should be checked or not.

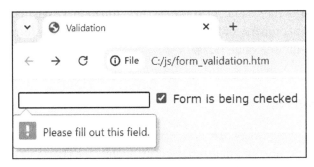

Figure 4.30 After Switching to noValidate = false

In the test2() function, the values of some Boolean properties of the element or the validity property are determined:

- willValidate: The element would be tested.
- validity.rangeOverflow and validity.rangeUnderflow: The value exceeds or falls below the valid range.
- validity.valid: The value would be sent as valid.

The value 82 in the second form leads to the result in Figure 4.31. In this case, the range-Overflow property of the validity object returns true.

Figure 4.31 Test Result for Value 82

For value 51, the result is as shown in Figure 4.32. Although the value is within the permitted range, it's odd and therefore not valid.

Figure 4.32 Test Result for Value 51

Each time you click the checkbox of the second form, the check2() function is executed. This sets the value of the disabled property of the form to true or false, and the form can only be operated if disabled has the value false.

4.10   Dynamically Created Forms

In this section, a form is created dynamically using JavaScript. Using a double for loop, the form then contains a larger number of input fields within a table (see Figure 4.33). The values of the id (for JavaScript) and name (for PHP) attributes and the event handlers are created using variables.

You can set the number of rows and columns in the program code at the beginning of the document. In this example, the table has five rows and three columns (i.e., fifteen input fields).

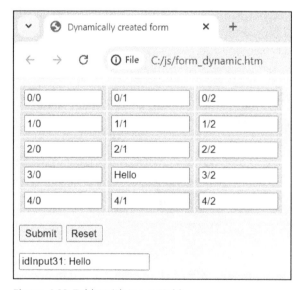

Figure 4.33 Table with Input Fields

The document code reads as follows:

```
... <head> ...
   <script>
      const rows = 5, columns = 3;

      function change(e)
      {
         const id = e.target.id;
         document.getElementById("idResponse").value
            = id + ": " + document.getElementById(id).value;
      }

      function submit()
      {
         let output = "";
         for(let r=0; r<rows; r++)
         {
            for(let c=0; c<columns; c++)
               output += document.getElementById("idInput"
                  + r + c).value + " ";
            output += "\n";
         }
         alert(output);
      }
   </script>
</head>
<body>
   <form id="idForm" method="post" action="form_dynamic.php">
   <table>
   <script>
      for(let r=0; r<rows; r++)
      {
         document.write("<tr>");
         for(let c=0; c<columns; c++)
         {
            document.write("<td><input id='idInput" + r + c + "' name='input"
               + r + c + "' value='" + r + "/" + c + "' size='10'></td>");
            document.getElementById("idInput" + r + c)
               .addEventListener("change", function(e) { change(e); });
         }
         document.write("</tr>");
      }

      document.getElementById("idForm").addEventListener("submit", submit);
```

```
        document.write("<input type='hidden' name='rows'"
           + " value='" + rows + "'>");
        document.write("<input type='hidden' name='columns'"
           + "value='" + columns + "'>");
     </script>
     </table>
     <p><input type="submit"> <input type="reset"></p>
     <p><input id="idResponse" readonly="readonly"></p>
     </form>
</body></html>
```

Listing 4.18 form_dynamic.htm File

The number of rows and columns is defined in the `rows` and `columns` variables at the beginning of the document.

The `r` and `c` loop variables serve as components for the values of the `id` and `name` attributes. Note the single quotation marks for the values of the attributes within the string for `document.write()`, which is in double quotation marks.

If it's determined that the content of an input field has changed, an anonymous function is called, and it calls the named function, `change()`. This determines the ID of the changed input field, which is output together with the new content of the input field.

In the `submit()` function, the contents of all input fields are determined and output, again using a double `for` loop.

You can use hidden form elements to send the number of rows and columns to the PHP file; in this way, you can dynamically determine the values sent. However, you should not exceed one thousand input fields, as otherwise, PHP will report an error in the default setting.

Chapter 5
The Document Object Model

The DOM provides read and write access to all elements of an HTML document.

The *Document Object Model* (DOM) is a model for accessing the elements of documents of certain types, such as HTML, XML (see Chapter 7, Section 7.3), and SVG (see Chapter 9).

The model is structured like a tree, similar to a directory tree, and it consists of individual nodes. There's a root node that has child nodes, which can in turn have child nodes, and so on. Each node can also have attribute nodes.

In JavaScript, each node is represented by a node object that can be an HTML element node, an HTML attribute node, or a text node. A node object can be read, changed, added, and deleted.

In JavaScript, you can access the content of the document via the document object, which provides methods with which you can create new document objects and access them. You have already frequently used the getElementById() method.

This chapter describes how you can access and change the elements of an HTML document, which is a prerequisite for many of the techniques covered in the following chapters.

5.1 Tree and Nodes

In this section, we want to take a look at the DOM tree of a simple HTML document. First, here's the HTML code:

```
<!DOCTYPE html><html lang="en-us">
<head>
   <meta charset="utf-8">
   <title>DOM, first example</title>
   <link rel="stylesheet" href="js5.css">
</head>
<body>
```

```
   <p>First paragraph</p>
   <p>Second paragraph with <span style="font-weight:bold;">
      bold font</span></p>
   <p>Third paragraph with <a href="embed.htm">a link</a></p>
</body>
</html>
```

Listing 5.1 dom_example.htm File

The document consists of three p containers, some of which contain further contain-ers. In the browser, it looks as shown in Figure 5.1.

Figure 5.1 Simple Example of DOM Tree

Modern browsers provide the option of visualizing the structure of the document according to the DOM. In the Google Chrome browser, you can do this via the devel-oper tools, which you have already learned about in Chapter 2, Section 2.5.5.

It's assumed that the browser is open with the mentioned program. Open the browser menu using the three dots in the top right-hand corner and select the **More tools •Developer tools** menu item. The developer tools are then displayed, and they offer you many different options.

Select the **Elements** tab in the upper area of the screen. As with a directory tree, you can now show or hide the individual levels (see Figure 5.2). If the tab for displaying the code is too small, you can drag the entire developer tools area to the left and thus give the individual tabs more space.

> **Note**
>
> In the following examples, we look at or change individual nodes at different levels. For the sake of clarity, you should always keep the document structure according to DOM in mind.

```
 ⫛  ⫐      Elements   Console   Sources   »      ⚙  ⋮  ✕
··· <!DOCTYPE html> == $0
 <html lang="en-us">
  ▼ <head>
     <meta charset="UTF-8">
     <title>DOM, first example</title>
     <link rel="stylesheet" href="js5.css">
  </head>
  ▼ <body>
     <p>First paragraph</p>
  ▼ <p>
     "Second paragraph with "
     <span style="font-weight:bold;"> bold font</span>
  </p>
  ▼ <p>
     "Third paragraph with "
     <a href="embed.htm">a link</a>
  </p>
  </body>
 </html>
```

Figure 5.2 Developer Tools, Elements Tab

5.2 Retrieving Nodes

In addition to the getElementById() method, you can use the getElementsByTagName() method. The parameter of the method is a string containing the name of the tag without angle brackets, and the method returns a field with references to all HTML elements with the desired tag.

The elements of the field are numbered with an index, starting at 0. You can access an individual field element via its index in rectangular brackets, and the length property of a field provides the number of elements. You can find out more about fields in Chapter 6, Section 6.1.

You can obtain the attribute nodes of an HTML element via the getAttribute() method of the relevant node object. The parameter of the function is a string containing the name of the attribute being searched for.

Here's a sample program:

```
... <head> ...
   <script>
      function retrieve()
      {
         let output = "";
         const parArray = document.getElementsByTagName("p");
         for (let i=0; i<parArray.length; i++)
```

```
                output += "Content: " + parArray[i].firstChild.nodeValue + "\n"
                   + "Attribute Style: " + parArray[i].getAttribute("style") + "\n";
             alert(output);
          }
      </script>
</head>
<body>
   <p style="font-family:Tahoma;">First paragraph</p>
   <p>Second paragraph</p>
   <p style="font-weight:bold;">Third paragraph</p>
   <form>
      <input type="button" id="idRetrieve" value="Retrieve">
   </form>
   <script>
      document.getElementById("idRetrieve").addEventListener("click", retrieve);
   </script>
</body></html>
```

Listing 5.2 dom_retrieve.htm File

The document contains three p containers in the HTML part, so the parArray field has three elements.

You can access the first child node of a node via the firstChild property of a node object. Each of the three paragraphs has only one text node as a child node.

The nodeValue property of a node object provides the associated text as a value. Calling the getAttribute("style") method returns the value of the style attribute of an HTML element node. As the second paragraph doesn't have this attribute, the method returns a *reference to nothing* (i.e., the value null).

The output of the program is shown in Figure 5.3. After you click the button, a dialog box appears with an output similar to the one shown in Figure 5.4.

Figure 5.3 Structure of Document with Three Paragraphs

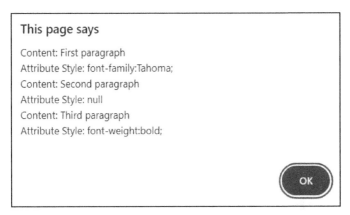

This page says

Content: First paragraph
Attribute Style: font-family:Tahoma;
Content: Second paragraph
Attribute Style: null
Content: Third paragraph
Attribute Style: font-weight:bold;

OK

Figure 5.4 Retrieving Nodes and Node Values

5.3   Child Nodes

Both a node and its child nodes are represented using a node object. The hasChild-Nodes() method returns the information on whether a node has child nodes or not, and the childNodes property of a node contains a field with the child nodes of a node.

The nodeType property contains the type of node. The value 1 stands for an element node (here, for an HTML element node), and the value 3 stands for a text node. The text content of a text node is in the nodeValue property, and the name of the tag of an HTML element node is in the nodeName property.

The following sample program runs through a part of the document's DOM tree using the properties and methods described earlier plus a recursive function:

```
... <head> ...
  <script>
    function nodes(reference, level)
    {
      for(let i=0; i<reference.childNodes.length; i++)
      {
        const ch = reference.childNodes[i];
        if(ch.nodeType == 3)
          document.write("Level:" + level + ", Text: "
            + ch.nodeValue + "<br>");
        else if(ch.nodeType == 1)
          document.write("Level:" + level + ", Tag: "
            + ch.nodeName + "<br>");
        if(ch.hasChildNodes())
        {
```

```
                level++;
                nodes(ch, level);
                level--;
            }
        }
    }
    </script>
</head>
<body>
    <p id="idParagraph">
        This <span style="font-style:italic;">is</span> a paragraph
        <span style="font-weight:bold;">with tags
        <span style="font-style:italic;">on multiple
        </span></span>levels</p>
    <script>
        const level = 0;
        nodes(document.getElementById("idParagraph"), level);
    </script>
</body></html>
```

Listing 5.3 dom_children.htm File

We are looking at the part of the DOM tree that contains the first paragraph, and the recursive nodes() function is called for the first time with a reference to this paragraph.

The level variable represents the current level within the tree. The child nodes of the paragraph are at level 0, their child nodes are at level 1 and their child nodes are at level 2.

In the nodes() function, the field with the child nodes of the current element is run through using a for loop, and a reference to the current field element is set up within the loop to shorten it. If the field element is a reference to a text node, its value will be output, but if it's a reference to an HTML element node, its name will be displayed.

If the current field element itself has child nodes, the number of the level is incremented and the nodes() function is called again.

In this way, you can run through all branches of the DOM tree. At some point, you'll determine that each of these branches has no child nodes, and the recursion will return. The number of the level will be reduced again, and at the end, you'll return to level 0.

The output of the program is shown in Figure 5.5.

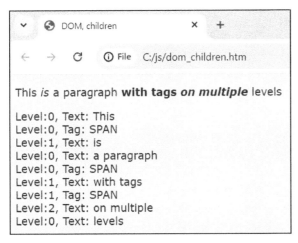

Figure 5.5 Child Nodes of First Paragraph

5.4   Adding Nodes

In this section, we create text nodes, HTML element nodes, and attribute nodes and add them to a document. We use the following two methods of the document object:

- The createTextNode() method, which creates a text node. The parameter of the method is a string containing the text.
- The createElement() method, which creates an HTML element node. The parameter of the method is a string with the tag without angle brackets.

In addition, three methods are used for a node object:

- The appendChild() method, which adds a child node to a node, at the end of the child node field. The parameter of the method is a reference to the node that gets added.
- The setAttribute() method, which sets the value of an attribute node. If the attribute node doesn't yet exist, it will be created first. The parameters of the method are two strings containing the name and the new value of the attribute.
- The insertBefore() method, which adds a child node to a node before another child node. The parameters of the method are two references: a reference to the node that gets added and a reference to the node before which it will be inserted.

Here's the program:

```
...  <head> ...
  <script>
    function append()
    {
      const text = document.createTextNode("Appended");
      const par = document.createElement("p");
```

```
        par.appendChild(text);
        par.setAttribute("style", "font-weight:bold;");
        document.getElementById("idBody").appendChild(par);
      }

      function insert()
      {
        const text = document.createTextNode("Inserted");
        const par = document.createElement("p");
        par.appendChild(text);
        par.setAttribute("style", "font-style:italic;");
        document.getElementById("idBody").insertBefore(par,
            document.getElementById("idParagraph3"));
      }
   </script>
</head>
<body id="idBody">
   <p id="idParagraph1">First paragraph</p>
   <p id="idParagraph2">Second paragraph</p>
   <p id="idParagraph3">Third paragraph</p>
   <p><input id="idAppend" type="button" value="Append">
      <input id="idInsert" type="button" value="Insert"></p>

   <script>
      document.getElementById("idAppend")
         .addEventListener("click", append);
      document.getElementById("idInsert")
         .addEventListener("click", insert);
   </script>
</body></html>
```

Listing 5.4 dom_add.htm File

In Figure 5.6, you can see the original document.

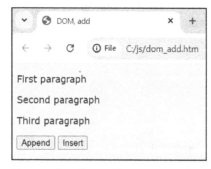

Figure 5.6 Before Adding

The **Append** and **Insert** buttons are used to append or insert a new HTML element node.

In the JavaScript function `append()`, a new text node with the text `Appended` is created, and an HTML element node is then created for a paragraph. The text node and an attribute node are added to the HTML element node, and then, the HTML element node is appended to the `body` node of the document at the end.

In the `insert()` function, a newly created HTML element node including the text node and attribute node is inserted before the specified child node of the `body` node.

In Figure 5.7, you can see the document after both buttons have been clicked twice.

Figure 5.7 After Adding

5.5   Changing Nodes

In this section, I will show you how you can change text nodes and HTML element nodes. Two further methods of the `node` object are introduced here:

- The `cloneNode()` method, which creates a copy of a node, optionally with or without all child nodes, their child nodes, and so on. The parameter of the method is a truth value. If this truth value has the value `true`, the node will be copied together with its entire child node structure.

- The `replaceChild()` method, which replaces the child node of a node with another child node. The parameters of the method are two references: one to the node that is being added and one to the node that is being replaced.

The `innerHTML` property of the `element` object is also very useful. It's used to retrieve or assign an HTML tag including text.

Here's the program:

```
... <head> ...
   <script>
      function textChange(id)
      {
         const par = document.getElementById(id);
         if(par.hasChildNodes())
            par.firstChild.nodeValue = "Text changed";
         else
         {
            const text = document.createTextNode("Text created");
            par.appendChild(text);
         }
      }

      function htmlChange()
      {
         const par = document.getElementById("idParagraph4");
         par.innerHTML = "Change <i>in fourth</i> paragraph";
      }

      function surround()
      {
         const par = document.getElementById("idParagraph3");
         const italic = document.createElement("span");
         italic.setAttribute("style", "font-style:italic;");
         const replaced = par.replaceChild(italic, par.firstChild);
         italic.appendChild(replaced);
      }

      function clone()
      {
         const par = document.getElementById("idParagraph3");
         const copy = par.cloneNode(true);
         document.getElementById("idBody").appendChild(copy);
      }
   </script>
</head>
<body id="idBody">
   <p id="idParagraph1">First paragraph</p>
   <p id="idParagraph2"></p>
   <p id="idParagraph3"><span style="font-weight:bold;">Third</span> paragraph</p>
   <p id="idParagraph4">Fourth paragraph</p>
   <p><input id="idTextChange" type="button" value="Change text">
```

```
    <input id="idTextCreate" type="button" value="Create text">
    <input id="idHtml" type="button" value="Change text and HTML"></p>
  <p><input id="idSurround" type="button" value="Surround node with HTML">
    <input id="idClone" type="button" value="Clone node"></p>

  <script>
    document.getElementById("idTextChange").addEventListener(
       "click", function() {textChange("idParagraph1");});
    document.getElementById("idTextCreate").addEventListener(
       "click", function() {textChange("idParagraph2");});
    document.getElementById("idHtml").addEventListener("click", htmlChange);
    document.getElementById("idSurround").addEventListener("click", surround);
    document.getElementById("idClone").addEventListener("click", clone);
  </script>
</body></html>
```

Listing 5.5 dom_change.htm File

In Figure 5.8, you can see the original document.

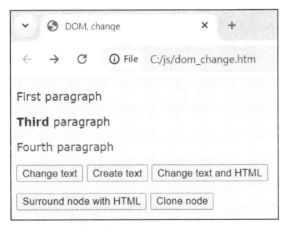

Figure 5.8 Before Changing

The second paragraph is not visible. Although it's created with a p container, it has no content (i.e., no child node).

The **Change text** and **Create text** buttons are used to change a text node. If the text node doesn't yet exist, it will be created first. The **Change text and HTML** button changes the content of a paragraph, and the **Surround node with HTML** button is used to surround a node (which may have child nodes) with an HTML node. The last button, **Clone node**, is used to copy a paragraph, including all child nodes. This button gets inserted at the end of the document.

The JavaScript textChange() function is called twice, each time with the unique ID of a paragraph as a parameter:

- The first time the function is called via the **Change text** button, and the first paragraph has a text node as its content. The hasChildNodes() method therefore returns true, and the content of the existing text node is replaced by other content. This also happens the second time the **Create text** button gets clicked, as the node in question now has a child node.

- The second time the function is called via the **Create text** button. This second paragraph has no content, and the hasChildNodes() method therefore returns false. A new text node is created and added to the paragraph as a new child node.

In the htmlChange() function, the content of the fourth paragraph is reset using the innerHTML property, namely, the HTML tag including text.

A new HTML element node is created in the surround() function. The new HTML element node replaces the first child node of the third paragraph, and this (former) first child node is then placed under the new node. This surrounds this child node with HTML code.

In the clone() function, a new node is created as a copy of the third paragraph, including all child nodes. This copy is appended to the end of the document.

In Figure 5.9, you can see the document after each button has been clicked once.

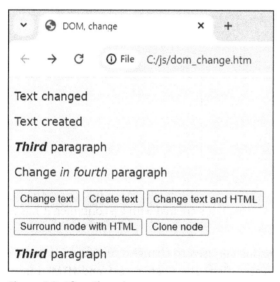

Figure 5.9 After Changing

5.6   Deleting Nodes

You can also delete nodes from the DOM tree, using the following methods for a node object:

- The removeChild() method, which deletes the child node of a node. The parameter of the method is a reference to the child node that gets deleted.

- The removeAttribute() method, which deletes the attribute node of a node. The parameter of the method is a string with the name of the attribute to be deleted.

Here's the program:

```
... <head> ...
  <script>
    function removeNode()
    {
      const par = document.getElementById("idParagraph2");
      document.getElementById("idBody").removeChild(par);
    }

    function removeAttribute()
    {
      const par = document.getElementById("idParagraph3");
      par.removeAttribute("style");
    }
  </script>
</head>
<body id="idBody">
  <p id="idParagraph1">First paragraph</p>
  <p id="idParagraph2">Second paragraph</p>
  <p id="idParagraph3" style="font-weight:bold;">Third paragraph</p>
  <p><input id="idNode" type="button" value="Remove node">
    <input id="idAttribute" type="button" value="Remove attribute"></p>

  <script>
    document.getElementById("idNode")
      .addEventListener("click", removeNode);
    document.getElementById("idAttribute")
      .addEventListener("click", removeAttribute);
  </script>
</body></html>
```

Listing 5.6 dom_delete.htm File

In Figure 5.10, you can see the original document.

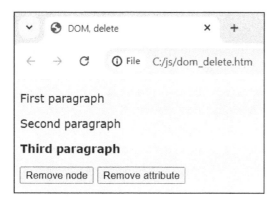

Figure 5.10 Before Deleting

The **Remove node** button deletes the second paragraph, and the **Remove attribute** button deletes the style attribute of the third paragraph.

In Figure 5.11, you can see the document after the deletions.

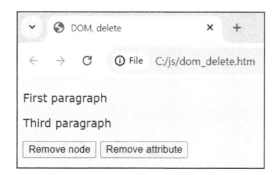

Figure 5.11 After Deleting

5.7   Creating a Table

In a final larger example, you can use a double for loop to create an entirely new table and embed it in the document. You use the createTextNode() and createElement() methods of the document object and the appendChild() method for a node object in this context. In the DOM tree, you first work from the outside in and then from the inside out again.

You create the table as follows:

- First, create the HTML element node for the table.
- The HTML element nodes for the rows follow within the outer loop.

- Within the inner loop, carry out the following steps, one after the other:
  - Create the text nodes for the cell.
  - Create the HTML element nodes.
  - Embed the text nodes in the HTML element nodes for the cells.
  - Embed the cells in the HTML element nodes for the rows.
- Within the outer loop, insert the HTML element nodes for the rows into the HTML element nodes for the table.
- Finally, embed the HTML element node for the table in the document.

Here's the program code:

```
... <head> ...
   <script>
      function createTable()
      {
         const table = document.createElement("table");

         for(let r=1; r<=3; r++)
         {
            const row = document.createElement("tr");
            for(let c=1; c<=5; c++)
            {
               const text = document.createTextNode("Cell " + r + "/" + c);
               const cell = document.createElement("td");
               cell.appendChild(text);
               row.appendChild(cell);
            }
            table.appendChild(row);
         }

         document.getElementById("idBody").appendChild(table);
      }
   </script>
</head>
<body id="idBody">
   <p><input id="idTable" type="button" value="Table"></p>
   <script>
      document.getElementById("idTable")
         .addEventListener("click", createTable);
   </script>
</body></html>
```

Listing 5.7 dom_table.htm File

After you click the **Table** button once, the document looks as shown in Figure 5.12.

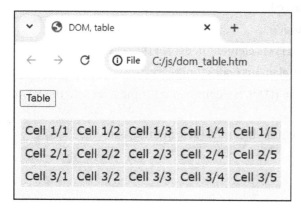

Figure 5.12 After Creating Table

Chapter 6
Using Standard Objects

In this chapter, you'll learn how to use arrays, strings, numbers, and time specifications.

Based on what you learned in Chapter 3, you can create your own objects. Very often, however, you also work in JavaScript with many predefined objects and their numerous properties and methods. This chapter describes the predefined objects—Array, String, Math, Number, and Date—that are used to edit arrays, strings, numbers, and mathematical problems as well as dates.

6.1   Arrays for Large Amounts of Data

You've already become familiar with the first simple arrays. They were simplified called fields. This section explains the topic in detail.

An *array* is used to store a group of thematically related data under a common name. Arrays consist of individual elements that are numbered, starting at 0. The number of an element is referred to as its *index*.

The values in an array usually have the same type (e.g., multiple strings as city names or several numbers as temperature values). However, you can also save values of different types in the same array.

In JavaScript, the Array object is used to create an array. It has properties and methods that enable you to access your own and predefined arrays.

6.1.1   One-Dimensional Arrays

In the following program, we'll create some one-dimensional arrays, run through them, and output them. Here's part 1:

```
...
<body><p>
   <script>
      let person = ["Ben", "Angela", "Will"];
      person = ["Peter", "Monica", "John"];
      const personTwo = person;
```

```
document.write("Simple output: " + personTwo + "<br><br>");

document.write("All elements:<br>");

document.write("for-Loop: ");
for(let i=0; i<person.length; i++)
   document.write(i + ":" + person[i] + " ");
document.write("<br>");

document.write("for-of-Loop: ");
for(let p of person)
   document.write(p + " ");
document.write("<br><br>");
```
. . .

Listing 6.1 array_one_dim.htm File, Part 1 of 4

Here, the array consists of a group of names. By default, an Array object is written within rectangular brackets and the individual elements are separated by commas.

The array is assigned to the person variable, which thus serves as a reference to the array. It was declared using let so another array can also be assigned to it.

You can create more references to an array, as we've done here by using personTwo, which is not a new array. The toString() method is already predefined for Array objects. It outputs the array elements one after the other, simply separated by commas.

The length property contains the length of an array (i.e., the number of elements). It can be used to control a for loop, for example.

There are two ways to access all elements of an array using a loop:

- Via a for loop. The first element of the array has the index 0 and is therefore called person[0], the second element has the index 1, the third has the index 2, and so on. The elements of the array can be changed within the for loop.

- Via a for-of loop, which was introduced with ECMAScript 2015. Each time the loop is run through, another element of the array gets assigned to a variable (in this case, the p variable). As this variable is a copy of an element, a change to this variable within the loop would not result in a change to the element.

In Figure 6.1, you can see the different outputs.

Let's now take a look at part 2 of the program:

. . .
```
document.write("Method at():<br>");

document.write("for-Loop: ");
```

```
for(let i=0; i<person.length; i++)
    document.write(i + ":" + person.at(i) + " ");
document.write("<br>");

document.write("Last elements: ");
document.write(person.at(-1) + " " + person.at(-2));
document.write("<br><br>");
```
...

Listing 6.2 array_one_dim.htm File, Part 2 of 4

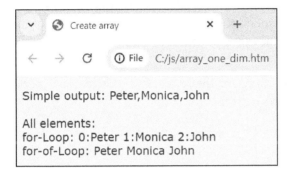

Figure 6.1 Different Outputs of Array

The at() method was introduced with ECMAScript 2022. This method accesses the element that has the index that is passed as a parameter. Unlike when you access the element using the square brackets, you can also pass a negative value. The –1 refers to the last element of the array, the –2 to the penultimate element, and so on.

You can see the outputs of part 2 in Figure 6.2.

```
Method at():
for-Loop: 0:Peter 1:Monica 2:John
Last elements: John Monica
```

Figure 6.2 at() Method

Part 3 of the program reads as follows:

...

```
document.write("Contains:<br>");
document.write("John: " + person.includes("John") + "<br>");
document.write("Julia: " + person.includes("Julia") + "<br><br>");

document.write("Convert, check:<br>");
const tx = "Hello world";
const tf = Array.from(tx);
```

```
document.write("Array from string: " + tf + "<br>");
document.write(tx + " is array: " + Array.isArray(tx) + "<br>");
document.write(tf + " is array: " + Array.isArray(tf) + "<br><br>");
...
```

Listing 6.3 array_one_dim.htm File, Part 3 of 4

The includes() method was introduced with ECMAScript 2016. It is called for an existing array, checks whether the array contains a specific value, and returns a truth value.

The Array.from() method is called for the Array class itself and not for an existing array, as there's no array at this point. For this reason, it's a *static method*, which creates a new array from a suitable object. Each element of the object becomes an element of the array, and in the case of a string, each individual character of the string becomes an element, as you can see in Figure 6.3.

The Array.isArray() method is also called for the Array class. This method checks whether an object is an Array object and returns a truth value.

```
Contains:
John: true
Julia: false

Convert, check:
Array from string: H,e,l,l,o, ,w,o,r,l,d
Hello world is array: false
H,e,l,l,o, ,w,o,r,l,d is array: true
```

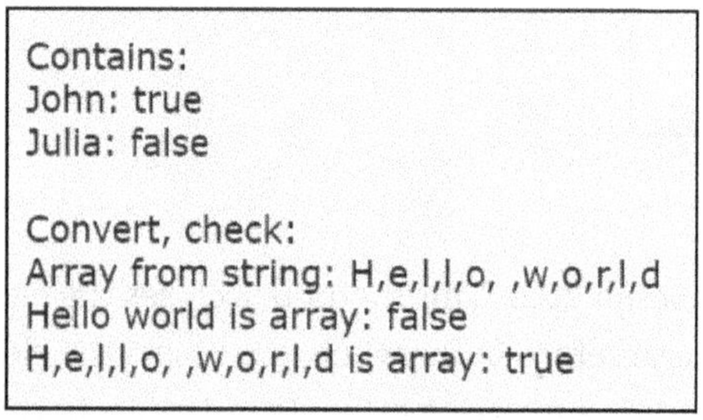

Figure 6.3 Checking and Converting

Now, here's part 4 of the program:

```
...
    document.write("Fill up:<br>");
    const za = new Array(7);
    document.write("At the beginning: " + za + "<br>");
    za.fill(42, 3, 5);
    document.write("Some elements: " + za + "<br>");
    document.write("Element 0: " + za[0] + "<br>");
    document.write("Element 7: " + za[7] + "<br>");
    za.fill(42, 3);
    document.write("At the end: " + za + "<br>");
    za.fill(42);
    document.write("All elements: " + za + "<br><br>");
  </script></p>
</body></html>
```

Listing 6.4 array_one_dim.htm File, Part 4 of 4

If you only know the size of an array but not its elements, you can create a new Array object using new. The desired size is indicated within the parentheses.

The fill() method is used to fill the array with a value, as follows:

- If the method is only called with one parameter, all array elements receive the specified value, which in this case is 42.

- An optional second parameter represents the index from which the array is filled with the specified value. which in this case is from index 3.

- An optional third parameter stands for the index up to which (excluded) the array is filled with the specified value, which in this case is from index 3 to before index 5.

If you access an element that has not yet been assigned a value, undefined will be output, in this case for the za[0] element. If you access an element outside the array, which here is za[7], undefined will be output as well. You can see the output in Figure 6.4.

```
Fill up:
At the beginning: ,,,,,,
Some elements: ,,,42,42,,
Element 0: undefined
Element 7: undefined
At the end: ,,,42,42,42,42
All elements: 42,42,42,42,42,42,42
```

Figure 6.4 Filling Array

> **Note**
>
> ECMAScript 2024 introduces the option of making arrays immutable using the # character. For example, let person = #["Ben", "Angela", "Will"]; means that the values of the array can no longer be changed. The person[0] = "Monica"; statement will then result in an error.

6.1.2   Multidimensional Arrays

You can create arrays that in turn contain arrays themselves. In this way, you create arrays with multiple dimensions. For example, the data of a table can be saved in a two-dimensional array, as you can see in part 1 of the following program:

```
...
<body><p>
   <script>
      document.write("Two-dimensional array:<br><br>");
      const za = [ [22.3, 18.5], [21.6, 19.7], [24.6, 20.1] ];

      document.write("Output with loop:<br>");
```

```
    for(let i=0; i<za.length; i++)
       for(let k=0; k<za[i].length; k++)
           document.write(i + "/" + k + ":" + za[i][k] + "<br>");
    document.write("<br>Simple output:<br>" + za + "<br>");

    document.write("<br>Method at():<br>");
    for(let i=0; i<za.length; i++)
       for(let k=0; k<za.at(i).length; k++)
           document.write(i + "/" + k + ":" + za.at(i).at(k) + "<br>");

    if(Array.prototype.flat)
    {
       const zb = za.flat();
       document.write("<br>After conversion, length: "
           + zb.length + "<br><br>");
    }
...
```

Listing 6.5 array_multi_dim.htm File, Part 1 of 2

The za array is an array that in turn consists of subarrays. Each subarray is enclosed in rectangular brackets, and all subarrays are again embedded in an array with rectangular brackets. I have inserted the spaces only for the purpose of clarification.

A two-dimensional array is created here, and the data of a small table is stored in it. For example, it shows the temperature values on three different days, measured in the morning and afternoon.

The length property of the entire array contains the number of subarrays, and the length property of a given subarray contains the number of elements of that subarray. You can use a double for loop with two different loop variables to output all elements of the array. These variables represent the index in the first and second dimension of the array.

Thanks to the predefined toString() method, multidimensional arrays can also be output easily. However, only the elements are listed, and the multidimensional structure of the array is then no longer recognizable, as you can see in Figure 6.5.

The at() method introduced in ECMAScript 2022 can also be used for multidimensional arrays. A chained call is used to output a single element: in the subarray that is accessed via at(i), the element that is accessed via at(k) gets accessed.

The flat() method was introduced with ECMAScript 2019. It can use internal recursion to create a one-dimensional array from the elements of a multidimensional array, and its number is output here.

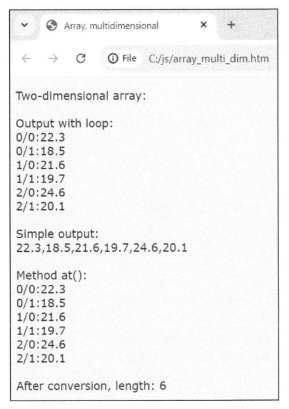

Figure 6.5 Two-Dimensional Array

Note

The objects in JavaScript have the prototype property. In the classic notation of object-oriented programming in JavaScript, this property stands for the prototype of an object, and it allows you to access this prototype. In this way, you can check whether a specific element exists for the browser you're using.

Note

The number of elements in the various subarrays could also be different.

Let's now take a look at part 2 of the program:

...

```
document.write("Three-dimensional array:<br>");
const zc = [[[19.8,17.4],[16.5,22.7]],
            [[15.1,16.9],[22.7,18.6]]];
```

```
          document.write("Output with loop:<br>");
          for(let i=0; i<zc.length; i++)
             for(let k=0; k<zc[i].length; k++)
                for(let m=0; m<zc[i][k].length; m++)
                   document.write(i + "/" + k + "/"
                      + m + ":" + zc[i][k][m] + "<br>");
          document.write("<br>Simple output:<br>" + zc + "<br>");

          if(Array.prototype.flat)
          {
             const zd = zc.flat(2);
             document.write("<br>After conversion with depth 2, length: "
                + zd.length);
          }
       </script></p>
</body></html>
```

Listing 6.6 array_multi_dim.htm File, Part 2 of 2

The zc array is three-dimensional. For example, the temperature values measured on different days, at different locations, and at different times could be stored in such an array.

A triple loop with three different loop variables is required to output the entire array. These variables represent the index in the first, second, and third dimension of the array. You can see the output in Figure 6.6.

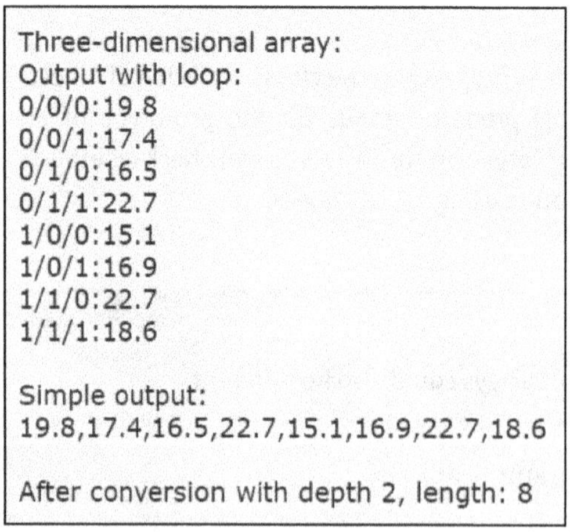

```
Three-dimensional array:
Output with loop:
0/0/0:19.8
0/0/1:17.4
0/1/0:16.5
0/1/1:22.7
1/0/0:15.1
1/0/1:16.9
1/1/0:22.7
1/1/1:18.6

Simple output:
19.8,17.4,16.5,22.7,15.1,16.9,22.7,18.6

After conversion with depth 2, length: 8
```

Figure 6.6 Three-Dimensional Array

You can call the flat() method with an optional value to specify the depth of the internal recursion up to which the elements of an array are collected. The default value for the depth is 1, but to collect all elements of a three-dimensional array, you must specify depth 2.

6.1.3   Arrays as Parameters and Return Values

An array can be passed to a function as a parameter. The transfer is made using the reference (*call by reference*). The values of the original array can therefore be changed in the function, and this applies regardless of whether the reference to the original array was assigned to a mutable variable or an immutable variable.

A function can return the reference to an array as a result. This holds true regardless of whether the array was previously passed as a parameter or whether it was only created in the function.

In the following program, these relationships are explained using a one-dimensional and a two-dimensional array. For the purpose of clarification, the definitions of the functions are written near the calls. Here's part 1 of the program:

```
...
<body><p>
   <script>
      function doubleOneDim(x)
      {
         for(let i=0; i<x.length; i++)
            x[i] *= 2;
      }

      document.write("Array as parameter:<br>");
      const za = [ 13, -2 ];
      document.write(za + "<br>");
      doubleOneDim(za);
      document.write(za + "<br>");
...
```

Listing 6.7 array_function.htm File, Part 1 of 4

The reference to a one-dimensional array is passed to the `doubleOneDim()` function. The elements of the array are changed in the function, and the array is output before and after the function is called (see Figure 6.7).

Here's part 2 of the program:

```
...
      function doubleTwoDim(x)
      {
         for(let i=0; i<x.length; i++)
            for(let k=0; k<x[i].length; k++)
```

```
            x[i][k] *= 2;
    }

    const zb = [ [5, 8], [6, -2] ];
    document.write(zb + "<br>");
    doubleTwoDim(zb);
    document.write(zb + "<br><br>");
...
```

Listing 6.8 array_function.htm File, Part 2 of 4

A two-dimensional array is passed to the doubleTwoDim() function. You can see the output in Figure 6.7.

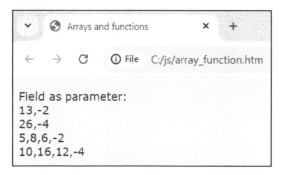

Arrays and functions   × +

← → C   ⓘ File   C:/js/array_function.htm

Field as parameter:
13,-2
26,-4
5,8,6,-2
10,16,12,-4

Figure 6.7 References to Arrays as Parameters

Here's part 3 of the program:

```
...
    function addOneDim(x, y)
    {
        const p = [];
        for(let i=0; i<x.length; i++)
            p[i] = x[i] + y[i];
        return p;
    }

    document.write("Array as return value:<br>");
    const zc = [ 5, 12 ];
    document.write(zc + "<br>");
    const zd = addOneDim(za, zc);
    document.write(zd + "<br>");
...
```

Listing 6.9 array_function.htm File, Part 3 of 4

Two one-dimensional arrays are passed to the addOneDim() function. The function uses empty square brackets to create an empty one-dimensional array, and each of the elements of this array is created by an assignment. The sum of the corresponding elements of the two transferred arrays is assigned.

The reference to the array is returned at the end of the addOneDim() function, and it's assigned to the zd variable. The output of this part of the program in shown in Figure 6.8.

Let's now look at part 4 of the program:

```
...
    function addTwoDim(x, y)
    {
        const p = [ [], [] ];
        for(let i=0; i<x.length; i++)
            for(let k=0; k<x[i].length; k++)
                p[i][k] = x[i][k] + y[i][k];
        return p;
    }

    const ze = [ [7, -4], [2, 9] ];
    document.write(ze + "<br>");
    const zf = addTwoDim(zb, ze);
    document.write(zf + "<br>");
    </script></p>
</body></html>
```

Listing 6.10 array_function.htm File, Part 4 of 4

Two two-dimensional arrays are passed to the addTwoDim() function, which uses several pairs of empty square brackets to create a two-dimensional array consisting of two empty one-dimensional arrays. The sum of the corresponding elements of the two transferred arrays is assigned.

The returned reference to the two-dimensional array is assigned to the zf variable, and you can see the output of this part of the program in Figure 6.8.

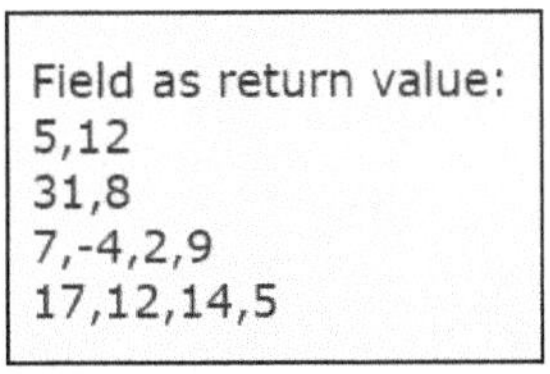

```
Field as return value:
5,12
31,8
7,-4,2,9
17,12,14,5
```

Figure 6.8 Reference to Array Is Returned

6.1.4 Callback Functions

You're already familiar with the term *callback function* from Chapter 2, Section 2.6.13. You can assign a reference to a callback function to a function, and the function then calls the passed callback function internally and uses its program logic.

There are some predefined methods for the Array object that use callback functions. In this example, some named callback functions are passed, and I have defined them close to their call for clarification. You can also pass anonymous callback functions.

Here's the program:

```
...
<body><p>
   <script>
      const za = [28.3, 16.5, 23.6, 16.7];
      document.write("Array: " + za + "<br><br>");

      document.write("Run with forEach():<br>");
      function output(element, index)
         { document.write(index + ":" + element + " "); }
      za.forEach(output);
      document.write("<br><br>");

      document.write("Examine with function:<br>");
      function over(value)
         { return value > this; }
      document.write("At least one value over 20: "
         + za.some(over, 20) + "<br>");
      document.write("All values over 20: "
         + za.every(over, 20) + "<br><br>");

      document.write("Filtering with function:<br>");
      const zb = za.filter(over, 20);
      document.write("New array with values over 20: " + zb + "<br><br>");

      document.write("Apply to function:<br>");
      function multiple(value)
         { return value * this; }
      const zc = za.map(multiple, 4);
      document.write("New array with values * 4: " + zc + "<br>");
   </script></p>
</body></html>
```

Listing 6.11 array_callback.htm File

The forEach() Array method is used to execute a callback function (in this case, the output() function) for all elements of an array. When the forEach() method is called, the name of the callback function gets transferred.

Internally, the forEach() method ensures that multiple parameters (some of which are optional) are passed to the output() callback function. These can appear as parameters in the definition of the output() callback function. The first parameter is a copy of the current element of the array, and the second parameter is optional and corresponds to the index of the current element. The output() callback function outputs all elements of the array with index and value.

The Array method some() checks whether at least one element of an array fulfills a certain condition within a callback function (here, over()). The Array method every() performs the same check for all elements of an array. Parameters (some of which are optional) are also passed internally to the over() callback function.

When calling one of the Array methods, you can also pass an additional parameter in addition to the name of the callback function. You can access this parameter within the callback function using this. The callback function over() then checks whether some (or all) elements have a value that is greater than the value of the additional transferred parameter.

The Array method filter() creates a new Array object, which contains copies of all elements of the original array that fulfill a certain condition within a callback function. Here, the callback function over() is called again and an additional parameter is also passed.

The Array method map() also creates a new Array object, which contains copies of all elements of the original array. The multiple() callback function is called here and multiplies the values of all elements by a factor that is passed as an additional parameter.

Here's an example of what calling the map() function with an anonymous callback function would look like: const zd = za.map(function(value) { return value * this; }, 4);. The first parameter of the map() function is not the name of the named function, but it is the complete definition of the function, without the function name.

The output of the program is shown in Figure 6.9.

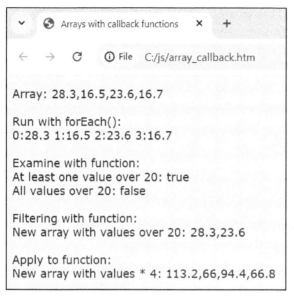

Array: 28.3,16.5,23.6,16.7

Run with forEach():
0:28.3 1:16.5 2:23.6 3:16.7

Examine with function:
At least one value over 20: true
All values over 20: false

Filtering with function:
New array with values over 20: 28.3,23.6

Apply to function:
New array with values * 4: 113.2,66,94.4,66.8

Figure 6.9 Callback Functions

6.1.5   Adding and Removing Elements

In JavaScript, arrays are dynamic. This section deals with methods of the Array object that enable you to add elements to an array or remove elements from an array.

The methods are as follows:

- The push() method, to add elements at the end of the array
- The pop() method, to remove elements from the end of the array
- The unshift() method, to add elements at the beginning of the array
- The shift() method, to remove elements from the beginning of the array
- The splice() method, to add or remove elements with regard to the middle of the array

The toSpliced() method has been available since ECMAScript 2023. It works like the splice() method, except that the result gets saved in a new array.

In addition to the methods mentioned here, the delete operator is used in the following sample program to delete the content of an element:

```
...
<body><p>
   <script>
      const city = ["Berlin", "Rome"];
      document.write("Array: " + city + "<br><br>");
```

```
    city.push("Bern", "Brussels");
    document.write("push: " + city + "<br>");

    const p = city.pop();
    document.write("pop: " + city + "<br>");
    document.write("Removed: " + p + "<br><br>");

    city.unshift("Madrid", "London");
    document.write("unshift: " + city + "<br>");

    const s = city.shift();
    document.write("shift: " + city + "<br>");
    document.write("Removed: " + s + "<br><br>");

    city.splice(2, 1, "Prague", "Sofia");
    document.write("splice to add: " + city + "<br>");

    city.splice(1, 2);
    document.write("splice to delete: " + city + "<br><br>");

    let cityTwo = city.toSpliced(1, 1, "Oslo", "Athens");
    document.write("add for new array: " + cityTwo + "<br>");

    cityTwo = city.toSpliced(1, 1);
    document.write("remove for new array: " + cityTwo + "<br><br>");

    delete city[1];
    document.write("delete: " + city);
  </script></p>
</body></html>
```

Listing 6.12 array_dynamic.htm File

An array with city names is created and changed several times. The array is output after each change, as you can see in Figure 6.10.

You can use the push() method to append one or more elements to the end, and you do this in the order in which the elements are specified as parameters when called. The pop() method removes an element from the end of the array and returns this element as the return value.

The unshift() and shift() methods do the same thing with regard to the start of the array. The shift() method returns the removed element as the return value.

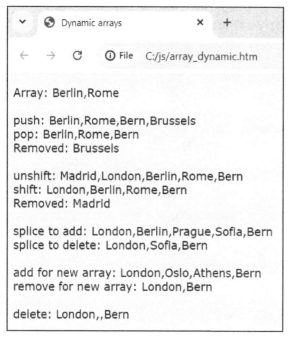

Figure 6.10 Adding and Removing Elements

The splice() method changes the array at any point. The first parameter specifies the position at which the removal or addition of elements begins, the second parameter determines how many elements are removed, and from the third parameter forward, you can specify one or more new elements to be added. If you don't specify a new element, only existing elements will be removed.

The toSpliced() method works like the splice() method, but the result is saved in a new array (here, in the cityTwo array).

The delete operator only deletes the content of an element, not the element itself. This creates a gap, so to avoid such a gap, you can use splice() instead of delete.

6.1.6   Changing Arrays

This section contains some methods that you can use to change arrays in their entirety. The result of the change can be saved within the same array or in a new array.

The methods for changing within an array are as follows:

- The concat() method, to connect two arrays
- The sort() method, to sort an array
- The reverse() method, to reverse the order of the elements of an array
- The join() method, to create a string from the elements of an array
- The copyWithin() method, to copy elements within an array

The methods (new since ECMAScript 2023) in which the result is saved in a new array without changing the original array are:

- toSorted(), for sorting an array
- toReversed(), to reverse the order of the elements of an array
- with(), to change a selected element of an array

Here's the program:

```
...
<body><p>
   <script>
      let city = ["Berlin", "Rome", "Oslo"];
      document.write("Original: " + city + "<br><br>");

      const cityTwo = ["Brussels", "Madrid", "London"];
      city = city.concat(cityTwo);
      document.write("Concatenated: " + city + "<br>");
      city.sort();
      document.write("Sorted: " + city + "<br>");
      city.reverse();
      document.write("Reversed: " + city + "<br><br>");

      const tx = city.join(" # ");
      document.write("Concatenated to a string: " + tx + "<br><br>");

      document.write("Previously: " + city + "<br>");
      city.copyWithin(1, 3, 5);
      document.write("Internally copied: " + city + "<br><br>");

      let cityThree = city.toSorted();
      document.write("Sorted into new array: " + cityThree + "<br>");
      cityThree = city.toReversed();
      document.write("Reversed into new array: " + cityThree + "<br>");
      cityThree = city.with(2, "Paris");
      document.write("With change into new array: " + cityThree);
   </script></p>
</body></html>
```

Listing 6.13 array_change.htm File

Again, an array with city names is created. Several methods are applied to this array, and Figure 6.11 shows the results.

The concat() method is called for the array created at the beginning, and the elements of another array are appended to this array in the specified order.

You can sort an array alphabetically using the sort() method. The order is based on the UTF-16 code, which is an encoding for Unicode characters. Among other things, upper-case letters have priority over lowercase letters, so the string oslo with a lowercase o would be located at the end. I describe sorting an array by numerical values in Section 6.1.8.

The reverse() method reverses the order of the elements. As the array was previously sorted, it's sorted in reverse in this case.

The join() method is used to convert an array into a string. The string contains all elements of the array, and each element is separated by the string that is specified as a parameter when join() gets called. For the reverse operation (i.e., the conversion of a string into an array), the split() method is available (Section 6.2.2).

The copyWithin() method can be called with up to three parameters. The first parameter specifies the position from which the copied elements overwrite the original elements. The elements copied there are selected as follows: the second parameter specifies the position of the first element, and the third parameter specifies the position after the last element. If the third parameter is omitted, all elements from the first element onwards are copied.

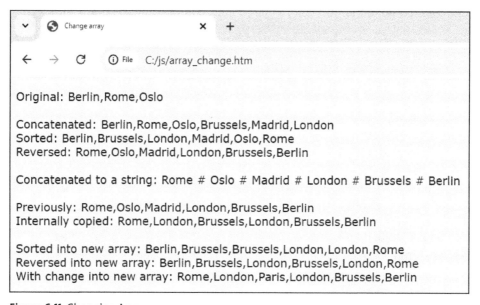

Figure 6.11 Changing Arrays

The toSorted() method is used to save the elements of an array sorted alphabetically in a new array. Accordingly, the toReversed() method is used to save the elements of an array reversed in a new array.

When the `with()` method is used, an array is saved in a new array, and this changes a selected element of the new array. The first parameter of the method specifies the position of this element, while the second parameter specifies its new value.

In all three cases, the original array doesn't get changed.

6.1.7 Copying Arrays

The `slice()` method is used to create a new array from specific (or all) elements of an existing array. Here's a sample program:

```
...
<body><p>
   <script>
      let city = ["Berlin", "Rome", "Oslo"];
      document.write("Original: " + city + "<br><br>");

      let cityTwo = city.slice();
      document.write("slice, from 0 to end, thus copy: " + cityTwo + "<br>");
      cityTwo = city.slice(2);
      document.write("slice, from 2 to end: " + cityTwo + "<br>");
      cityTwo = city.slice(-2);
      document.write("slice, from -2 to end: " + cityTwo + "<br>");
      cityTwo = city.slice(1, 2);
      document.write("slice, from 1 to 2 excluded: " + cityTwo + "<br>");
      cityTwo = city.slice(0, -2);
      document.write("slice, from 0 to -2 excluded: " + cityTwo + "<br><br>");
   </script></p>
</body></html>
```

Listing 6.14 array_copy.htm File

The already known array of city names serves as the original array. The `slice()` method is used to create a new array that contains specific (or all) elements of the original array, and the method has two optional parameters that are used to specify the position.

If the method is called without parameters, the entire original array will get copied to a new array. If the method is called with one parameter, all elements from the specified position onwards are copied to a new array. If a call is made with two parameters, the new array contains the elements from the first specified position to *before* the second specified position.

If one of the two parameters has a negative value, the corresponding position is calculated from the end of the original array. Remember, the index 0 indicates the first element and the index –1 indicates the last element.

The output of the program is shown in Figure 6.12.

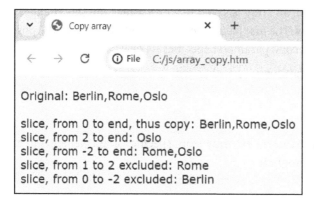

Figure 6.12 Copying Arrays

6.1.8   Sorting Number Arrays

Using a callback function, you can use the sort() method to sort arrays by number. Here's a sample program:

```
... <head> ...
   <script>
      function ascending(x, y)
      {
         if(x > y) return 1;
         else      return -1;
      }

      function descending(x, y)
      {
         if(y > x) return 1;
         else      return -1;
      }
   </script>
</head>
<body><p>
   <script>
      const za = [3, 28, 2, 14];
      document.write("Unsorted: " + za + "<br>");
      za.sort();
      document.write("Sorted by characters: " + za + "<br>");
      za.sort(ascending);
      document.write("Sorted by numbers, ascending: " + za + "<br>");
      za.sort(descending);
      document.write("Sorted by number, descending: " + za + "<br><br>");
```

```
      const zb = [3, 28, 2, 14];
      const zc = zb.toSorted(ascending);
      document.write("Sorted to a new array: " + zc);
   </script></p>
</body></html>
```

Listing 6.15 array_sort.htm File

The array of numbers named za is initially unsorted.

If the sort() method is called without parameters, the array will be sorted alphabetically, as described in Section 6.1.6. This means, for example, that the number 14 comes before the number 2. This is because the digits are considered individually and the digit 1 comes before the digit 2—but that doesn't make any sense for a number array.

When calling the sort() method, you can pass the reference to a comparison function to be used for sorting. In the first case, you want to sort by number in ascending order. Here, a reference to the ascending() function is passed to the method, and this function is used internally by the sort() method to compare two elements.

If the first element is larger than the second element in this comparison, a positive value is returned; otherwise, a negative value is returned. This results in the two compared elements being sorted in ascending order. This is carried out repeatedly for two elements during the internal sorting process of the sort() method, so at the end, all elements of the array are sorted in ascending order.

The descending() function is passed for sorting by numbers in descending order. In Figure 6.13, you can see the result of the program.

You can also pass the reference to a comparison function to the toSorted() method when calling it. This function was introduced in ECMAScript 2023. It sorts the elements of an array by number in ascending order and saves them in a new array. The original array isn't changed in this process.

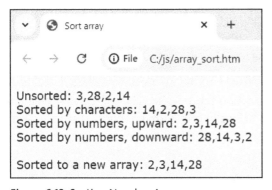

Figure 6.13 Sorting Number Array

207

6.1.9   Finding Elements in an Array

The find() and findIndex() methods also work with callback functions that have access to internally transferred parameters. The methods find the first element of an array that meets a certain condition or its index, and in each case, another parameter that can contribute to the search can be passed.

ECMAScript 2023 introduced the closely related findLast() and findLastIndex() methods. They are used to find the last element that fulfills a certain condition or its index.

Here's a sample program:

```
...
<body><p>
   <script>
      const za = [3, 14, 2, 28, 5, 8];
      document.write("Array: " + za + "<br>");

      function elementGreaterValue(element)
         { if(element > this) return element; }
      document.write("First element > 10: "
         + za.find(elementGreaterValue, 10) + "<br>");
      document.write("Last element > 10: "
         + za.findLast(elementGreaterValue, 10) + "<br>");

      function indexGreaterValue(element, index)
         { if(element > this) return index; }
      document.write("Index for first element > 10: "
         + za.findIndex(indexGreaterValue, 10) + "<br>");
      document.write("Index for last element > 10: "
         + za.findLastIndex(indexGreaterValue, 10) + "<br>");
   </script></p>
</body></html>
```

Listing 6.16 array_find.htm File

An array of numbers named za is created.

When the find() method is called, a reference to the elementGreaterValue() callback function is transferred. This callback function returns the value of the first element that is greater than the value of the additionally transferred parameter. All elements are checked one after the other, and as soon as a suitable element is found, the check is aborted and the element is returned.

When the findIndex() method is called, a reference to the indexGreaterValue() callback function is transferred. It returns the index of the first value that fulfills the check, and Figure 6.14 shows the results.

The findLast() and findLastIndex() methods work in the same way, in relation to the last element or its index.

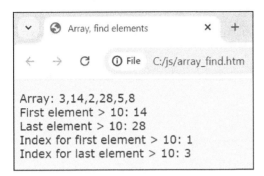

Figure 6.14 Finding Elements in Array

6.1.10 Destructuring and the Spread Operator

In Chapter 2, Section 2.6.6, you learned about the *destructuring assignment*. Due to destructuring and the *spread operator*, ... (three dots), there are further possibilities for arrays, which are explained next on the basis of examples.

Here's part 1 of the program:

```
...
<body><p>
   <script>
      let a, b, c, remainder;
      [a, b, c] = [12, 15, -6];
      document.write("Variables: " + a + " " + b + " " + c + "<br>");
      [a, b, c] = [12, 15];
      document.write("Too few array elements: "
         + a + " " + b + " " + c + "<br>");
      [a, b] = [12, 15, -6];
      document.write("Too many array elements: " + a + " " + b + "<br>");

      [a, ...remainder] = [12, 15, -6];
      document.write("Variable: " + a + ", Remainder to array: "
         + remainder + "<br>");

      const f = [3, 28, 4];
      [a, b, c] = f;
      document.write("From array: " + a + " " + b + " " + c + "<br><br>");
...
```

Listing 6.17 array_destructuring.htm File, Part 1 of 3

The array [12, 15, –6] is assigned to the three variables in a single statement. If there are too few variables, the remaining variables are assigned the undefined value. If there are too many variables, the array elements expire.

Using the spread operator ..., you can make sure that surplus array elements are assigned to another array (here, remainder).

All operations have the same behavior if there's a variable on the right-hand side of the assignment that references an array.

The output of this part of the program is shown in Figure 6.15.

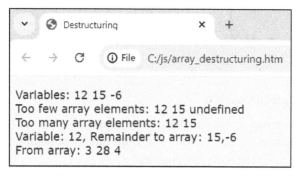

Figure 6.15 Destructuring and Spread Operator

Here's part 2 of the program:

```
...
    function add(a, b, ...remainder)
    {
        let sum = a + b;
        for(let i=0; i<remainder.length; i++)
            sum += remainder[i];
        return sum;
    }

    document.write("One number: " + add(3) + "<br>");
    document.write("Two numbers: " + add(3, 5) + "<br>");
    document.write("Six numbers: "
        + add(3, 5, 4, 9, 3, 2) + "<br><br>");
...
```

Listing 6.18 array_destructuring.htm File, Part 2 of 3

You can also use the spread operator . . . to make a function more flexible. You can pass two or more parameters to the add() function, which adds up all parameters. The first two parameters are stored in the a and b variables, while all other parameters are stored in the remainder array. If there are less than two parameters, an error will occur.

The output of this part of the program is shown in Figure 6.16.

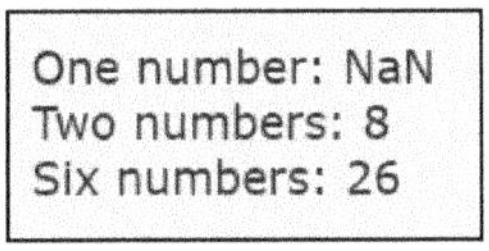

```
One number: NaN
Two numbers: 8
Six numbers: 26
```

Figure 6.16 Function and Spread Operator

Here's part 3 of the program:

```
...
    function maxi(x)
    {
       let maximum = x[0][0], row = 0, column = 0;
       for(let i=0; i<x.length; i++)
          for(let k=0; k<x[i].length; k++)
             if(x[i][k] > maximum)
             {
                maximum = x[i][k];
                row = i;
                column = k;
             }
       return [maximum, row, column];
    }

    const zb = [ [-3, 28, 4], [6, -5, 12] ];
    document.write("Array: " + zb + "<br>");
    let maximum, row, column;
    [maximum, row, column] = maxi(zb);
    document.write("Maximum:" + maximum + ", Row:"
       + row + ", Column:" + column + "<br>");
  </script></p>
</body></html>
```

Listing 6.19 array_destructuring.htm File, Part 3 of 3

Destructuring allows a function to return more than one value. The program determines the maximum value of a two-dimensional array and the first position of its occurrence in the array, which consists of the index of the row and the index of the column. The program returns all three values.

In the maxi() function, the two-dimensional array is transferred to the variable x. First, it's specified that the largest value is in the top-left array element (i.e., at row index 0 and column index 0). All array elements are checked, and if an array element is found that is greater than the previous maximum, its value and position are saved. At the end of the function, an array with the three results is returned and it's assigned to three variables.

The output of this part of the program is shown in Figure 6.17.

```
Array: -3,28,4,6,-5,12
Maximum:28, Row:0, Column:1
```

Figure 6.17 Destructuring of Function Return

6.1.11   Arrays of Objects

You can group together the references to objects within an array. Thanks to the toString() method, the output of such an array is simple. Such an array can be sorted using suitable sorting criteria.

Here's a sample program:

```
... <head> ...
   <script>
      class Car
      {
         constructor(c, s)
         {
            this.color = c;
            this.speed = s;
         }

         toString()
         {
            return this.color + "/" + this.speed;
         }
      }

      function descending(x, y)
      {
         if(y.speed > x.speed)
            return 1;
         else
            return -1;
      }
   </script>
</head>
<body><p>
   <script>
```

```
        const vc = [new Car("Red", 50),
            new Car("Blue", 85), new Car("Yellow", 65)];
        document.write("unsorted: " + vc + "<br>");
        vc.sort(descending);
        document.write("sorted: " + vc);
    </script></p>
</body></html>
```

Listing 6.20 array_object.htm File

The program creates three Car objects and saves the references to these objects in an array named vc. The toString() method is available for both Array objects and Cars, so the entire array can be output using the document.write() method.

The array is sorted in descending order of speed, and when the sort() method is called, the name of a callback function that makes this possible gets transferred. The two elements of the array that are compared internally are references to objects of the Car class. By comparing the two values of the speed property, the callback function returns a positive or negative value., which ensures that the array will be sorted correctly. The result is shown in Figure 6.18.

Figure 6.18 Arrays of Objects with Sorting

> **Note**
>
> You can find the u_one_dimensional and u_multi_dimensional exercises in bonus chapter 1, sections 1.15 and 1.16, in the downloadable materials for this book at *www.rheinwerk-computing.com/5875*.

6.2   Processing Strings

You've already become familiar with strings, which are similar to arrays and are used to store texts that originate from input, for example. The individual characters of a string are numbered, starting at 0, and the number of a single character in a string is also referred to as the index.

The String object with its properties and methods can be used to read and change strings. In some examples, you've already seen the + operator for concatenating strings and the combined assignment operator += for extending a string.

6.2.1   Creating and Checking Strings

The following program creates a string and determines its length. ECMAScript 2015 introduced the includes(), startsWith(), and endsWith() methods, and you can use these methods to determine whether a string contains, starts with, or ends with another string.

Here's the program:

```
...
<body><p>
    <script>
        const tx = "info@rheinwerk-publishing.com";
        document.write("Text: " + tx + "<br>");
        document.write("Length: " + tx.length + "<br>");
        document.write("Length: " + "publish".length + "<br>");

        document.write("Contains 'publish': " + tx.includes("publish") + "<br>");
        document.write("Contains 'Publish': " + tx.includes("Publish") + "<br>");
        document.write("Contains 'publish' from 16: "
            + tx.includes("publish", 16) + "<br>");

        document.write("Starts with 'info': " + tx.startsWith("info") + "<br>");
        document.write("Starts with 'info' from 1: "
            + tx.startsWith("info", 1) + "<br>");

        document.write("Ends with 'com': " + tx.endsWith("com") + "<br>");
        document.write("Ends with 'com' before 28: "
            + tx.endsWith("com", 28) + "<br><br>");
    </script></p>
</body></html>
```

Listing 6.21 string_check.htm File

The tx variable is assigned a string with an email address, and the length property contains the number of characters within the string. Properties and methods can be applied not only to a variable that references a String object but also directly to a character string that is in double quotation marks (see "publish".length).

Up to two parameters are passed to the includes() method. The first parameter corresponds to the string that is searched for in the String object, and the second, optional parameter can be used to specify the position in the String object from which the search starts.

The situation is similar with the startsWith() method. A second parameter can also be specified for the endsWith() method, and in this case, the String object is first shortened to the specified length for the search. The program then checks whether it ends with the string you're looking for.

The output of the program is shown in Figure 6.19.

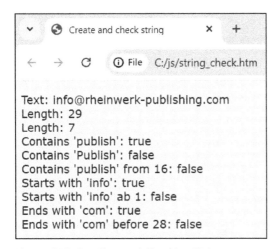

Figure 6.19 Creating and Checking String

6.2.2   Elements of a String

As with an array, you can determine the individual characters of a string by using square brackets. In addition, the at() method has been available for this purpose since ECMAScript 2022, also with a negative index.

You can use the split() method to split a string using a separator string. An Array object is created with the individual parts of the string, and in this context, the join() method of the Array object is shown once again, and it in turn joins the array into a character string.

The fromCodePoint() method, which was introduced with ECMAScript 2015, creates a string from a sequence of *code points*, which correspond to the number of a character in the UTF-16 code. The codePointAt() method returns this number for a single character within a string.

Here's a program with examples:

```
...
<body><p>
   <script>
      const tx = "info@rheinwerk-publishing.com";
```

```
        document.write("Text: " + tx + "<br>");
        document.write("Characters with []: ");
        for(let i=0; i<5; i++)
            document.write(i + ":" + tx[i] + " ");
        document.write("...<br>Characters with at(): ");
        for(let i=0; i<5; i++)
            document.write(i + ":" + tx.at(i) + " ");
        document.write("...<br>");

        document.write("Last elements: ");
        document.write(tx.at(-1) + " " + tx.at(-2));
        document.write("<br><br>");

        document.write("Split in array: ");
        const a = tx.split(".");
        for(let i=0; i<a.length; i++)
            document.write(i + ":" + a[i] + " ");
        document.write("<br>");
        const txNew = a.join(".");
        document.write("Join array to string: " + txNew + "<br><br>");

        const uc = String.fromCodePoint(97, 48, 65,
                    228, 246, 252, 196, 214, 220, 223);
        document.write("Codepoints result in string: " + uc);
        document.write("<br>String results in codepoints:<br>");
        for(let i=0; i<uc.length; i++)
            document.write(i + ":" + uc.codePointAt(i) + " ");
        document.write("<br><br>");
    </script></p>
</body></html>
```

Listing 6.22 string_elements.htm File

The individual characters are output together with the index in two different ways, using two for loops.

The string is then split using the split() method based on the dots (i.e., using the "." separator string). The individual array elements don't contain the separator string itself. Using the Array method join() and the same separator string, a new string is created from the array elements.

The String.fromCodePoint() method is called for the String object itself and not for an existing string, as there's no string at this point. Such a method is referred to as a *static method* (see also Section 6.1.1).

The lowercase letters have code points 97 to 122, the uppercase letters have code points 65 to 90, and the numbers have code points 48 to 57. The German umlauts and the ß character, for example, are outside this regular sequence. The codePointAt() method in turn returns the code points.

You can see the output in Figure 6.20.

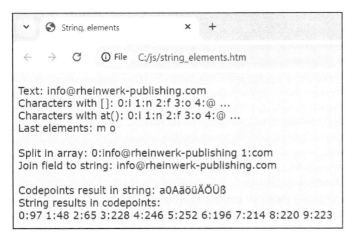

Figure 6.20 Elements of String

6.2.3 Searches and Substrings

The indexOf() and lastIndexOf() methods return the index at the first or last position at which a string begins within another string. If the string is not found, the value –1 will be returned.

The substring() and substr() methods return the copy of a part of the string. Here's a program with examples:

```
...
<body><p>
   <script>
      const tx = "info@rheinwerk-publishing.com";
      document.write("Text: " + tx + "<br>");
      document.write("First character 'i': " + tx.indexOf("i") + "<br>");
      document.write("Last character 'i': " + tx.lastIndexOf("i") + "<br>");

      document.write("All characters 'i': ");
      let pos = tx.indexOf("i");
      while pos != -1:
      {
         document.write(pos + " ");
         pos = tx.indexOf("i", pos+1);
      }
```

```
        document.write("<br><br>");

        document.write("From 4 to end: " + tx.substring(4) + "<br>");
        document.write("From 4 to 7 excluded: " + tx.substring(4,7) + "<br>");
        document.write("From 4, length 6: " + tx.substr(4,6) + "<br><br>");
    </script></p>
</body></html>
```

Listing 6.23 string_position.htm File

First, the program determines the position of the first or last character i within the string.

In the indexOf() method, you can optionally specify the position from which the search should start. Accordingly, the lastIndexOf() method can optionally specify the position up to which the search should continue.

You can use a while loop to search for the positions of all characters i. The search for the next occurrence of the character i always starts at the position where the last occurrence had previously been found.

If you specify only one parameter each for the substring() and substr() methods, the part of the string that starts at the specified position and extends to the end of the string will be returned. The second, optional parameter in the substring() method specifies which character is to be the first to no longer be copied. In contrast, you can use the substr() method to specify the number of characters to be copied.

You can see the output in Figure 6.21.

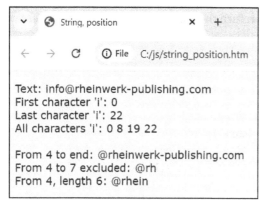

Figure 6.21 Search and Substrings

6.2.4 Changing Strings

There are a number of methods available for changing a string. The toUpperCase() and toLowerCase() methods change each letter to an uppercase letter or a lowercase letter.

The `replace()` method replaces the first occurrence of a searched string with another string. ECMAScript 2021 introduced the `replaceAll()` method, which performs the same task for all occurrences of a searched string and is already used by all modern browsers.

The `repeat()` method was introduced with ECMAScript 2015, and it returns the desired number of repetitions of a string.

Since ECMAScript 2017, the `padStart()` and `padEnd()` methods have been available. You fill a string at the beginning or end with another, possibly repeated string, up to a desired length.

The `trim()` method was introduced with ECMAScript 2015. It removes all *whitespace characters*—for example, spaces, tabs, or the characters for the end of a line— at the beginning and end of a string. Since ECMAScript 2019, there have also been the `trimStart()` and `trimEnd()` methods, which only remove the named characters at the beginning or end.

Here's a program with examples:

```
...
<body><p>
  <script>
    const tx = "info@rheinwerk-publishing.com";
    document.write("Text: " + tx + "<br>");
    const tg = tx.toUpperCase();
    document.write("All uppercase: " + tg + "<br>");
    document.write("All lowercase again: " + tg.toLowerCase() + "<br><br>");

    document.write("Replace: " + tx.replace("info", "mail") + "<br>");
    document.write("Replace all: " + tx.replaceAll("i", "I") + "<br><br>");

    const ha = "Hello";
    document.write("Repeat: " + ha.repeat(5) + "<br><br>");
    document.write("Padding at the beginning: "
        + ha.padStart(10,"xy") + "<br>");
    document.write("Padding at the end: " + ha.padEnd(10,"xy") + "<br><br>");

    const tr = " \t Hello\t \n \r";
    document.write("Before trimming: |" + tr + "|<br>");
    document.write("Codepoints: ");
    for(let i=0; i<tr.length; i++)
       document.write(tr.codePointAt(i) + " ");
    document.write("<br>");
    document.write("After trimming at the beginning: |"
        + tr.trimStart() + "|<br>");
```

```
        document.write("After trimming at the end: |" + tr.trimEnd() + "|<br>");
        document.write("After trimming: |" + tr.trim() + "|<br>");
    </script></p>
</body></html>
```

Listing 6.24 string_change.htm File

The tg variable is assigned a copy of the tx string, which only contains uppercase letters. The tx string itself remains unchanged, and a copy of tg is then output containing only lowercase letters.

In the string, the first (and in this case only) occurrence of the string "info" is replaced by "mail". Then, all i characters are replaced by the I character.

The "Hello" string is returned in five repetitions, and then, it's filled with up to ten characters using the "xy" string, once at the beginning and once at the end. The "xy" string is repeated twice completely and once incompletely.

Another string contains a total of thirteen characters before trimming. There are some whitespace characters at the beginning and end: space (code point 32), tabulator (code point 9), and the characters for the end of the line (code point 10 or 13). After trimming at the beginning, there are only ten characters left, after trimming at the end, there are only eight characters left, and after all the trimming, there are only 5 characters left. For clarification, the | character is displayed before and after the string.

You can see the output in Figure 6.22.

Figure 6.22 Changing Strings

6.2.5   Checking a Password

As an application example for the String object, the check of an entered password is carried out. If it doesn't meet the criteria in Figure 6.23, the corresponding form will not be submitted.

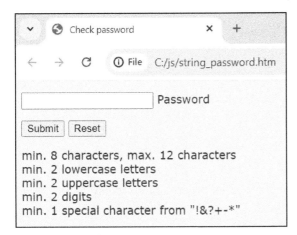

Figure 6.23 Checking Password

Here's the program:

```
... <head> ...
  <script>
     const lowercase = "abcdefghijklmnopqrstuvwxyz";
     const uppercase = "ABCDEFGHIJKLMNOPQRSTUVWXYZ";
     const number = "0123456789";
     const special = "!&?+-*";

     function submit(e)
     {
        const pw = document.getElementById("idPassword").value;
        if(pw.length < 8 || pw.length > 12)
        {
           alert("Password must contain 8-12 characters");
           e.preventDefault();
           return false;
        }

        let l = 0, u = 0, n = 0, s = 0;
        for(let i = 0; i < pw.length; i++)
        {
           if(lowercase.includes(pw[i])) l++;
```

```
            if(uppercase.includes(pw[i])) u++;
            if(number.includes(pw[i])) n++;
            if(special.includes(pw[i])) s++;
        }
        if(l < 2 || u < 2 || n < 2 || s < 1)
        {
            alert("Minimum number not met");
            e.preventDefault();
            return false;
        }
        return true;
    }
  </script>
</head>
<body>
  <form id="idForm" method="post" action="string_password.php">
    <p><input id="idPassword" type="password" name="password"> Password</p>
    <p><input type="submit"> <input type="reset"></p>
    <p>min. 8 characters, max. 12 characters<br>
      min. 2 lowercase letters<br>
      min. 2 uppercase letters<br>
      min. 2 digits<br>
      min. 1 special character from "!&?+-*"</p>
  </form>
  <script>
    document.getElementById("idForm").addEventListener
        ("submit", function(e) {return submit(e);});
  </script>
</body></html>
```

Listing 6.25 string_password.htm File

Some strings are saved with the following permitted characters:

- The lowercase letters and the uppercase letters, without any language-specific special characters
- The numbers from 0 to 9
- A selection of special characters

When the form is submitted, the submit() function gets called. The password entered is saved as a string. If it's too short or too long, the corresponding error message appears, the function terminates, and the form will not be submitted.

If the string is of the correct length, some counters are first declared and set to 0. The individual characters of the input are then run through using a `for` loop, and the `includes()` method is used to determine whether they are one of the permitted characters. If they are, the corresponding counter will be increased.

After the loop, it's determined whether any of the counter values don't meet the criteria for the respective minimum number. If it doesn't, a corresponding error message will also be displayed here, the function will terminate, and the form will not be submitted.

If all criteria are met, the form gets submitted.

> **Note**
>
> Bonus chapter 2 in the downloadable materials for this book (at *www.rheinwerk-computing.com/5875*) deals with the `regexp` object. It also provides further options for processing strings using regular expressions.

6.2.6 Unicode Characters

The Unicode character set comprises a very large number of symbols and characters from many different languages and character systems. Since ECMAScript 2024, you've been able to ensure that Unicode characters are well-formed (i.e., displayed uniformly and correctly in the browser).

The code of a Unicode character consists of two parts within a string. Each part consists of the character combination \u and four hexadecimal digits.

In addition, you can use the `isWellFormed()` string method to check whether a string corresponds to a well-formed code for a Unicode character.

In the following program, some characters from the field of emojis are shown, and then some tests are carried out:

```
...
<body><p>
    <script>
        document.write("\uD83D\uDE00" + "\uD83D\uDE0F" + "<br>");
        document.write("\uD83D\uDE10" + "\uD83D\uDE1F" + "<br>");
        document.write("\uD83D\uDE20" + "\uD83D\uDE2F" + "<br>");
        document.write("\uD83D\uDE30" + "\uD83D\uDE3F" + "<br>");
        document.write("\uD83D\uDE40" + "\uD83D\uDE4F" + "<br>");

        document.write("Hello".isWellFormed() + "<br>");
        document.write("\uDE00".isWellFormed() + "<br>");
        document.write("\uD83D\uDE00".isWellFormed() + "<br>");
```

```
    </script></p>
</body></html>
```

Listing 6.26 character_unicode.htm File

You can see the result of the program in Figure 6.24.

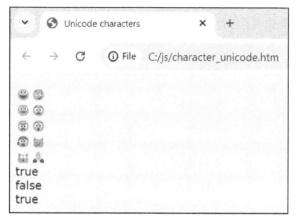

Figure 6.24 Unicode Characters

6.3   Numbers and Math

The predefined Math and Number objects make it easier for you to work with numbers and mathematical calculations. In this section, we'll explain many functions that you're familiar with from the calculator, and you'll also learn about simple random number generators as well as those that are suitable for encryption.

6.3.1   Math Object

The Math object allows you to use a range of useful mathematical methods, and it also provides some mathematical constants as properties.

Let's first take a look at the sample program:

```
...
<body><p>
    <script>
        const a = 4.75, b = -4.75;
        let output = "a:" + a + " b:" + b + "<br>"
            + "ceil(a):" + Math.ceil(a) + "   ceil(b):" + Math.ceil(b) + "<br>"
            + "floor(a):" + Math.floor(a)
            + "   floor(b):" + Math.floor(b) + "<br>"
            + "round(a):" + Math.round(a)
```

```
      + "   round(b):" + Math.round(b) + "<br>"
      + "abs(a):" + Math.abs(a) + "   abs(b):" + Math.abs(b) + "<br>"
      + "sign(a):" + Math.sign(a) + "   sign(b):" + Math.sign(b)
      + "   sign(0):" + Math.sign(0) + "<br>"
      + "max(a,b):" + Math.max(a,b)
      + "   min(a,b):" + Math.min(a,b) + "<br><br>";

   const c = 64;
   output += "c:" + c + "   sqrt(c):" + Math.sqrt(c)
      + "   cbrt(c):" + Math.cbrt(c) + "<br>"
      + "pi:" + Math.PI + "   e:" + Math.E + "<br>"
      + "exp(4.6):" + Math.exp(4.6)
      + "   log(100):" + Math.log(100) + "<br>"
      + "pow(10,3):" + Math.pow(10,3)
      + "   log10(1000):" + Math.log10(1000);
   document.write(output);
  </script></p>
</body></html>
```

Listing 6.27 number_math.htm File

First, a positive number and a negative number are rounded to integers using the ceil(), floor(), and round() rounding methods. The ceil() method rounds a number to the next largest integer, and the floor() method rounds a number to the next smallest integer.

The round() method rounds a number to an integer in accordance with commercial practice. For example, with positive numbers, 3.000 to 3.499 become 3 and 3.500 to 3.999 become 4. With negative numbers, –3.000 to –3.499 become –3 and –3.500 to –3.999 become –4.

The abs() method calculates the absolute value of a number (i.e., the number without its sign). The sign() method has been available since ECMAScript 2015, and it determines the sign of a number and returns the value 1 for a positive number, –1 for a negative number, and 0 for the number 0. The max() and min() methods return the largest or smallest number of any given set of numbers.

The sqrt() method determines the square root of a number, while the cbrt() method determines the cube root. The PI and E properties contain the mathematical constant for Pi (π) or Euler's number e (i.e., the base of the exponential function).

The exp() method calculates the result of the exponential function (i.e., e to the power of). Here, the method determines e to the power of 4.6, which is approximately 100. The log() method calculates the natural logarithm of a number (i.e., the logarithm of a number to the base e), which is mathematically referred to as ln. This corresponds to the inverse of the exp() method. In the example, the method determines the natural logarithm of 100 (*ln 100*), which is approximately 4.6.

The pow() method determines the result of the power function (i.e., *x to the power of y*). Here, the method determines 10 to the power of 3, which is 1,000. The log10() method has been available since ECMAScript 2015, and it calculates the logarithm of a number to the base 10. Here, the method determines the logarithm of 1,000, which is 3.

You can see the result of the program in Figure 6.25.

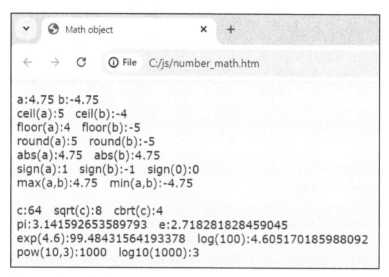

Figure 6.25 Math Object, Methods, and Properties

6.3.2   Trigonometric Functions

The following program is used to display a table with the values of the sine, cosine, and tangent trigonometric functions for angles from 0 to 90 degrees, in steps of 15 degrees.

The sin(), cos(), and tan() methods of the Math object calculate the values based on an angle in radians. The 360 degrees of a full circle correspond to 2π in radians, so an angle in degrees must be converted into radians using the factor $\pi/180$.

Here's the corresponding sample program:

```
...
<body>
    <table>
        <tr>
            <td style="font-weight:bold;">Angle</td>
            <td style="font-weight:bold;">Radian</td>
            <td style="font-weight:bold;">Math.sin()</td>
```

```
            <td style="font-weight:bold;">Math.cos()</td>
            <td style="font-weight:bold;">Math.tan()</td>
        </tr>
    <script>
        for(let i=0; i<=90; i+=15)
        {
            const rad = i / 180 * Math.PI;
            document.write("<tr><td>" + i + "</td><td>"
                + rad.toFixed(3) + "</td><td>"
                + Math.sin(rad).toFixed(3) + "</td><td>"
                + Math.cos(rad).toFixed(3) + "</td><td>"
                + Math.tan(rad).toFixed(3) + "</td></tr>");
        }
    </script>
    </table>
</body></html>
```

Listing 6.28 number_math_angle.htm File

The first step consists of the conversion to radians. The results are rounded to three decimal places using the toFixed() method of the Number object, but unfortunately, the conversion from 90 degrees (= $\pi/2$) to radians is not mathematically exact, so the tangent of 90 degrees doesn't have a value of infinity.

The output of the program is shown in Figure 6.26.

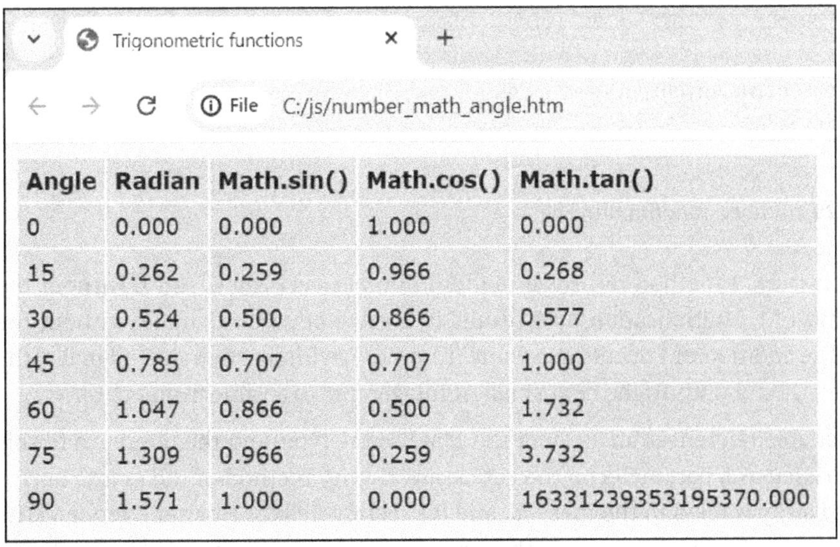

Angle	Radian	Math.sin()	Math.cos()	Math.tan()
0	0.000	0.000	1.000	0.000
15	0.262	0.259	0.966	0.268
30	0.524	0.500	0.866	0.577
45	0.785	0.707	0.707	1.000
60	1.047	0.866	0.500	1.732
75	1.309	0.966	0.259	3.732
90	1.571	1.000	0.000	16331239353195370.000

Figure 6.26 Table with Trigonometric Functions

6.3.3   Random Generators and Typed Arrays

The random() method of the Math object has already been used frequently in this book to determine simple random numbers. Here, it's used to roll dice ten times.

The crypto object is used to determine better random numbers that can also be used for encryption purposes. The getRandomValues() method of the crypto object fills a typed array with random numbers. What are *typed arrays*? By default, variables in Java-Script do not have data types and can store any values (e.g., integers, numbers with decimal places, strings, truth values). In addition, the need has grown to provide storage areas with a fixed size for storing numbers. *Typed arrays* were created for this purpose.

In the following program, a typed array of size 10 of the Uint32Array type is filled with random numbers:

```
...
<body><p>
   <script>
      let output = "random(): ";
      for(let i=0; i<10; i++)
         output += Math.floor(Math.random() * 6 + 1) + " ";

      output += "<br><br>crypto():";
      const za = new Uint32Array(10);
      crypto.getRandomValues(za);
      za.sort();
      for(let i=0; i<za.length; i++)
         output += "<br>" + za[i];

      document.write(output);
   </script></p>
</body></html>
```

Listing 6.29 number_random.htm File

The Math.random() method returns a random number between 0 and 1, without the upper limit of 1. Multiplication by 6 results in a value between 0 and 6, without the value 6. The addition of 1 results in a value from 1 to 7, without the value 7. Finally, the floor() method rounds to the next smallest integer (i.e., to a value from 1 to 6).

The za variable references an array of the Uint32Array type, and the reference to this array is passed as a parameter to the getRandomValues() method of the crypto object. The typed array is filled in the method, and it's then sorted for a better overview. The sort() method of a typed array sorts the array according to numerical values, and after that, the array gets displayed.

For example, the output of the program can look like the one shown in Figure 6.27.

Figure 6.27 Random Generators and Typed Arrays

There are different types of typed arrays, both for integers and for numbers with decimal places. At this point, only the `Int32Array` and `Uint32Array` types are described. Using these types, you can store integers in a storage area with a size of 32 bits = 4 bytes.

A number from −2,147,483,648 to +2,147,483,647 can be stored in an element of an `Int32Array`. The `U` stands for *unsigned*, and only positive numbers from 0 to +4,294,967,295 can be stored in an element of a `Uint32Array`.

6.3.4   Integers

The `Number` object provides properties and methods related to the precision and the size of the number range in JavaScript. In addition, I present the `BigInt` object for very large integers. Here's a sample program:

```
...
<body><p>
   <script>
      let nu = Number.MAX_SAFE_INTEGER;
      document.write("Largest safe integer: " + nu + "<br>");
      document.write("Safe integer: "
         + Number.isSafeInteger(nu) + "<br>");
      nu = nu + 1;
      document.write("+1: Safe integer: "
         + Number.isSafeInteger(nu) + "<br>");
      nu = Number.MIN_SAFE_INTEGER;
      document.write("Smallest safe integer: " + nu + "<br><br>");
```

```
    nu = Number.MAX_SAFE_INTEGER;
    x = nu + 1;
    y = nu + 2;
    document.write("Error without BigInt:<br>" + x + "<br>" + y + "<br><br>");

    const numberBig = BigInt(nu);
    x = numberBig + 1n;
    y = numberBig + 2n;
    document.write("No error with BigInt:<br>" + x + "<br>" + y);
  </script></p>
</body></html>
```

Listing 6.30 number_integer.htm File

ECMAScript 2015 introduced the MAX_SAFE_INTEGER and MIN_SAFE_INTEGER properties of the Number object. They correspond to the largest or smallest *safe integer*. Only a safe integer can be used for mathematically exact calculations in JavaScript. A number greater than MAX_SAFE_INTEGER or less than MIN_SAFE_INTEGER is saved as a number with decimal places. Both the storage and the comparison of integers is only mathematically exact for safe integers, and you can use the isSafeInteger() method to check whether a specific number is a safe integer.

Since ECMAScript 2020, however, there has been another option. The BigInt object works with even larger integers, and a number of the BigInt type is generated either by appending the character n or by calling the BigInt() function. In this program, you can see that two different integers that are greater than MAX_SAFE_INTEGER are only correctly distinguished as BigInt values.

You can see the result of the program in Figure 6.28.

Figure 6.28 Integers

6.3.5   Numbers with Decimal Places

This section describes the properties and methods of the Number object for numbers with decimal places, and it also introduces typed arrays for this group of numbers.

Let's first take a look at a sample program:

```
...
<body><p>
   <script>
      let nu = 1000 / 7;
      document.write("Number: " + nu + "<br>");
      document.write("Display with 0 decimal places: "
         + nu.toFixed(0) + "<br>");
      document.write("Display with 2 decimal places: "
         + nu.toFixed(2) + "<br>");
      document.write("As exponential number: "
         + nu.toExponential() + "<br><br>");

      nu = Number.MAX_VALUE;
      document.write("Number with the highest value: " + nu + "<br>");
      nu = nu * 10;
      document.write("Enlarged: " + nu + "<br>");

      nu = Number.MIN_VALUE;
      document.write("Number with the lowest value: " + nu + "<br>");
      nu = nu / 10;
      document.write("Reduced: " + nu + "<br><br>");

      nu = Number.EPSILON;
      document.write("&epsilon; = Smallest distance: " + nu + "<br>");
      nu = 1 + Number.EPSILON;
      document.write("1 + &epsilon;: " + nu.toFixed(40) + "<br>");
      nu = 1 + Number.EPSILON / 2;
      document.write("1+ &epsilon;/2: " + nu.toFixed(40) + "<br><br>");

      let ar = new Float32Array(2);
      ar[0] = 1000 / 7;
      document.write("Float32: " + ar[0].toFixed(20) + "<br>");

      ar = new Float64Array(2);
      ar[0] = 1000 / 7;
      document.write("Float64: " + ar[0].toFixed(20) + "<br>");
```

```
    </script></p>
</body></html>
```

Listing 6.31 number_point.htm File

The program calculates the value of 1,000 / 7, which is 142.857142857142.... In standard JavaScript variables, numbers can only be stored with 14 to 15 decimal places, so the result becomes 142.85714285714286.

The toFixed() method of the Number object returns a string containing a number that is rounded to a specified amount of decimal places. The number itself doesn't get changed in this context. The amount 0 results in a string that contains an integer, and with the amount 2, the number is rounded to two decimal places.

The toExponential() method of the Number object results in an output of the number as an exponential number with a base and an exponent. The number displayed here is pronounced as follows: 1.42857... times 10 to the power of 2.

Both methods are applied directly to the variables without naming the Number object.

The MAX_VALUE property corresponds to the number with the largest value that can be processed using JavaScript. If you increase it, you get a value of infinity.

On the other hand, the MIN_VALUE property corresponds to the number with the smallest value that can be processed using JavaScript. If you reduce it, the value is 0.

The EPSILON property was introduced with ECMAScript 2015, and it represents the difference between the value 1 and the smallest number with decimal places that is greater than 1.

Note

The term *epsilon* is used in mathematics to describe the smallest possible distance between two values. A smaller distance can't be detected in JavaScript.

In *typed arrays* of the Float32Array or Float64Array type (see also Section 6.3.3), numbers with decimal places are stored in a memory area of size 32 bits = 4 bytes or 64 bits = 8 bytes. The former provides accuracy to 6 to 7 decimal places, while the latter provides accuracy to 14 to 15 decimal places.

You can see the result of the program in Figure 6.29.

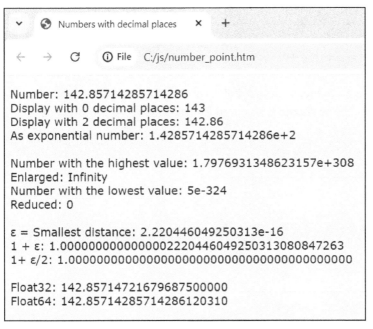

Figure 6.29 Numbers with Decimal Places

6.3.6   Custom Extension for Numbers

In this section, we expand the options for outputting numbers. For this purpose, we extend the Number object by a custom format() method and use it to fill a number with leading characters for output.

You can call the method with one or two parameters. The first parameter stands for the desired total number of digits in the output, and you can pass a string as the second parameter, which serves as a padding character. If there's no second parameter, it's filled with zeros.

The method is defined in an external file named *number.js*. You can fill this file with other useful methods for numbers, and after integrating the file, you can use those methods.

Here's the program that uses the method from the external file:

```
... <head> ...
   <script src="number.js"></script>
</head>
<body>
   <script>
      const x = 15;
      document.write(x.format(5) + "<br>");
      document.write(x.format(5, "#"));
```

```
   </script>
</body></html>
```

Listing 6.32 number_extend.htm File

The program assigns a number with two digits. For the output, it's filled up to five digits. For the first output, the character 0 is used for padding; for the second output, the hash character #.

You can see the output in Figure 6.30.

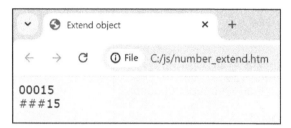

Figure 6.30 Custom format() Method

This is followed by the external file in which the format() method is defined:

```
Number.prototype.format = function(number_of_digits, character = "0")
{
   let formatted = this.toString();
   while(formatted.length < number_of_digits)
      formatted = character + formatted;
   return formatted;
}
```

Listing 6.33 number.js File

The objects in JavaScript have the prototype property. In the classic notation of object-oriented programming in JavaScript, this property stands for the prototype of an object and allows you to extend all objects that have the same prototype. We use this property for the Number object in this program.

An anonymous function is assigned to the format property of the Number.prototype object, and this turns format() into a method.

You can transfer one or two parameters. The first parameter specifies the desired number of digits, and the second parameter corresponds to the padding character, and it has the character 0 as the default value. A different padding character can be passed as the second parameter during the call.

You can access the Number object for which the method is called via this. The content of the object (i.e., the number) is converted into a string using the toString() method.

As long as the length of this string is less than the required number of digits, the padding character is added at the beginning. The filled string then gets returned.

6.4   Using Time Specifications

The Date object provides numerous options for calculating and outputting date and time information.

The point of origin for the Date object is January 1, 1970, at 00:00 Greenwich Mean Time (GMT), although today, we use Coordinated Universal Time (UTC). Date objects had negative values before that point in time and positive values afterward. The times are calculated in milliseconds.

6.4.1   Creating Time Data

In this section, you'll see various options for creating a new Date object:

- Without parameters, a Date object is created with the system time of the PC (i.e., with the current date and time).
- With one parameter, the value is taken for the milliseconds since the point of origin (January 1, 1970 at 00:00).
- With three parameters, the values are for the year, month, and day.
- With six parameters, the values are for the year, month, day, hour, minute, and second.

The numbering of months starts at 0, which is why there are values from 0 (January) to 11 (December). If you enter a parameter with a value that is too large or too small, it will automatically be converted (e.g., January 32 becomes February 1).

Here's the program:

```
...
<body><p>
   <script>
      const timeA = new Date();
      document.write("Current date: " + timeA + "<br>");

      const timeB = new Date(2024, 6, 10);
      document.write("Date from day, month, year: " + timeB + "<br>");

      const timeC = new Date(2024, 6, 10, 17, 8, 3);
      document.write("Hour, minute, second in addition: " + timeC + "<br>");
```

```
        const timeD = new Date(2024, 1, 30, 17, 68, -3);
        document.write("From values for conversion: " + timeD + "<br>");

        const timeE = new Date(2 * 60 * 1000);
        document.write("Date from milliseconds: " + timeE + "<br>");

        const timeF = new Date(-1.5 * 24 * 60 * 60 * 1000);
        document.write("Negative value: " + timeF);
    </script></p>
</body></html>
```

Listing 6.34 time_create.htm File

The result is shown in Figure 6.31.

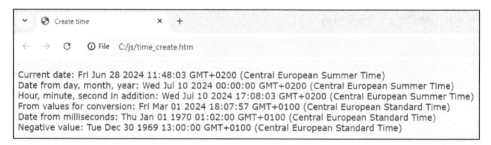

Figure 6.31 Creating Time Data

The Date objects created are output in the default format, which contains the abbreviated English name of the day of the week and the month, as well as the day of the month, the year, and the time. Further output options can be found in Section 6.4.2.

GMT+0100 means that the local time is one hour ahead of GMT, due to a time difference of one hour compared to London-Greenwich. In the case of a further time difference of one hour due to daylight saving time, GMT+0200 would appear. The names *Central European Standard Time* (CET) and *Central European Summer Time* (CEST) are also displayed.

The other time specifications are generated using transferred values. The entries 2024, 6, 10 become July 10, 2024, and the time is set to 00:00:00 or 17:08:03.

Here are some values to convert: February 30, 2024, becomes March 1, 2024, and 17:68 becomes 18:08. If we subtract 3 seconds, it becomes 18:07:57.

The value 2 × 60 × 1,000 milliseconds corresponds to a difference of two minutes from the point of origin. The result is 01.01.1970 00:02 GMT, and due to the time difference, this corresponds to 01:02 for CET.

The negative value –1.5 × 24 × 60 × 60 × 1,000 corresponds to a difference of 1.5 days, and the result is 30.12.1969 12:00 GMT. Due to the time difference, this corresponds to 13:00 for CET.

6.4.2 Outputting Time Data

In this section, you'll get to know various options for outputting a Date object. I will explain this using the current period as an example:

- You already know the output of the object without a method.
- The toLocaleString() method generates country-specific information that is obtained from the information on the browser currently in use.
- The toUTCString() method generates the UTC. After adding one hour, the result is CET.
- The getTimezoneOffset() method returns the time offset of the local time compared to UTC in minutes. The state of South Australia, for example, has a time difference of 9.5 hours.
- The dateFormat() method is a custom extension of the Date object for the formatted output of a time. A number of predefined methods are used internally (Section 6.4.3).

When calling the dateFormat() method, you can pass a string with format specifications that provide the following information:

- w: weekday
- d: day of the month, two digits
- M: number of the month, two digits
- y: year, four digits
- h: hour, two digits
- m: minute, two digits
- s: second, two digits
- i: milliseconds

The two-digit values are output with a leading zero if necessary. Other characters, such as : (colon) or . (dot), are not converted but are simply output at the relevant position.

Here's the program:

```
...  <head> ...
   <script src="number.js"></script>
   <script src="date.js"></script>
</head>
<body><p>
   <script>
```

```
    const time = new Date();
    const output = "Time object: " + time + "<br>"
        + "Local format: " + time.toLocaleString() + "<br>"
        + "UTC format: " + time.toUTCString() + "<br>"
        + "Local offset from UTC: "
        + (time.getTimezoneOffset()/60) + " hour(s)<br>"
        + "Default format: " + time.dateFormat("M/d/y h:m:s") + "<br>"
        + "With weekday: " + time.dateFormat("w, M/d/y") + "<br>"
        + "With milliseconds: " + time.dateFormat("h:m:s,i") + "<br>";
    document.write(output);
  </script></p>
</body></html>
```

Listing 6.35 time_output.htm File

Two extensions are integrated: first, the extension of the Number object from the *number.js* file (Section 6.3.5), and second, the extension of the Date object from the *date.js* file (Section 6.4.3). Internally, the dateFormat() method uses the format() method from the *number.js* file to output the leading zeros.

When you use the toLocaleString() and toUTCString() methods, the name of the day of the week and the name of the month are also listed. The toUTCString() method specifies the time in UTC.

The getTimezoneOffset() method returns an indication in minutes, which are converted into hours here. Three different outputs follow as examples of dateFormat().

The result is shown in Figure 6.32.

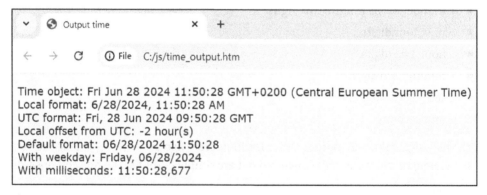

Figure 6.32 Outputting Time Data

6.4.3   Extension of the Date Object

The separate dateFormat() method is used to extend the Date object and provides a convenient way to output different time specifications.

First, here's the code:

```
Date.prototype.dateFormat = function(formatcharacter)
{
   const wday = ["Sunday", "Monday", "Tuesday",
      "Wednesday", "Thursday", "Friday", "Saturday"];
   let output = "";

   for(let i=0; i<formatcharacter.length; i++)
      switch(formatcharacter[i])
      {
         case "w": output += wday[this.getDay()];            break;
         case "d": output += this.getDate().format(2);       break;
         case "M": output += (this.getMonth()+1).format(2);  break;
         case "y": output += this.getFullYear();             break;
         case "h": output += this.getHours().format(2);      break;
         case "m": output += this.getMinutes().format(2);    break;
         case "s": output += this.getSeconds().format(2);    break;
         case "i": output += this.getMilliseconds();         break;
         default:  output += formatcharacter[i];
      }
   return output;
}
```

Listing 6.36 date.js File

An anonymous function is assigned to the dateFormat property of the Date.prototype object, and this turns dateFormat() into a method. You pass a string with the required format details as a parameter, and this string is run through character by character in a for loop to assemble and return the formatted time specification.

In preparation, an array is created with the weekdays spelled out. The getDay() method of the Date object returns the number of the day of the week: the value 0 represents Sunday, the value 1 represents Monday, and so on up to the value 6 for Saturday. These values are used as an index for the array.

You can access the Date object for which the method is called via this. Within the for loop, each individual character is examined using a switch branch. The different format characters lead to calls of various methods of the Date object: getDay(), getDate(), get-Month(), getFullYear(), getHours(), getMinutes(), getSeconds(), and getMilliseconds(). There are a few special features you should note:

- The getMonth() method returns a value between 0 and 11, so the value 1 must be added for the output.

- For some values, the format() method of the Number object is called to output a leading zero.

If a character doesn't correspond to any of the format characters, it gets immediately added to the formatted time specification without further conversion. This applies, for example, to colons, dots, or spaces.

6.4.4   Calculating with Time Data

You can also use the Date object to offset times against each other. The following program contains four examples:

- Calculating the time difference between two times
- Calculating the duration of a program section
- Calculating the age of a person
- Adding a time difference to a time specification using the custom dateAdd() method of the Date object

The last method is defined in the external *date.js* file, just like the separate dateFormat() method.

Here's the first part of the program:

```
...  <head> ...
   <script src="number.js"></script>
   <script src="date.js"></script>
</head>
<body><p>
   <script>
      let timeA = new Date(2024, 6, 15, 23, 59, 20);
      let timeB = new Date(2024, 6, 16, 0, 1, 50);
      document.write("1: " + timeA.dateFormat("M/d/y h:m:s") + "<br>");
      document.write("2: " + timeB.dateFormat("M/d/y h:m:s") + "<br>");
      let difference = timeB - timeA;
      document.write("Difference in minutes: "
         + (difference / 60000) + "<br><br>");
...
```

Listing 6.37 time_compute.htm File, Part 1 of 4

Two Date objects are created, each with predefined data for date and time. To calculate the date and time difference between them, you can simply subtract one from another. A time specification in a Date object is stored internally as the number of milliseconds since the point of origin (i.e., since January 1, 1970, 00:00). Consequently, the difference is a number as well. To represent it in minutes, divide it by 60,000.

The result is shown in Figure 6.33.

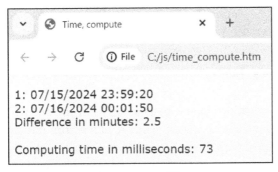

Figure 6.33 Calculating Time Differences

Here's the second part of the program:

```
...
    const msecA = new Date();
    let a = 1;
    for(let i=0; i<5e7; i++)
        a = a * -1;
    const msecB = new Date();
    difference = msecB - msecA;
      document.write("Computing time in milliseconds: "
        + difference + "<br><br>");
...
```

Listing 6.38 time_compute.htm File, Part 2 of 4

The program queries the current system time twice, once before a program section and once after. In this way, the program calculates the amount of time the computer needs to process a program section in which a `for` loop is run very frequently. The period depends on the performance and the current utilization of the computer.

You can see this result in Figure 6.33 as well.

Let's now look at the third part of the program:

```
...
    timeA = new Date(1979, 7, 16);
    timeB = new Date();
    difference = timeB.getFullYear() - timeA.getFullYear();
     if(( timeB.getMonth() < timeA.getMonth() )
          || ( timeB.getMonth() == timeA.getMonth()
          && timeB.getDate() < timeA.getDate() ) )
       difference--;
    document.write("1: " + timeA.dateFormat("M/d/y") + "<br>");
```

```
        document.write("2: " + timeB.dateFormat("M/d/y") + "<br>");
        document.write("Age: " + difference + "<br><br>");
...
```

Listing 6.39 time_compute.htm File, Part 3 of 4

The program generates two time specifications, one with a date of birth and one with the current system time. It calculates the age as follows: The age of a person initially corresponds to the difference in years between the two times. Then the program compares the months from the two times, and if the month of birth has not yet been reached or if it's currently the month of birth and the birthday has not yet been reached, the program deducts the value 1.

> **Note**
>
> The parentheses within the linked condition are not necessary and are for clarification purposes only. The logical AND operator takes precedence over the logical OR operator.

The result is shown in Figure 6.34.

```
1: 08/16/1979
2: 06/28/2024
Age: 44

1: 07/31/2024 17:55:30
3 days, 6 hours and 10 minutes later:
2: 08/04/2024 00:05:30
```

Figure 6.34 Calculating Age and Adding Time Specification

The fourth and final part of the program reads as follows:

```
...
        timeA = new Date(2024, 6, 31, 17, 55, 30);
        timeB = timeA.dateAdd(0, 0, 3, 6, 10, 0);
        document.write("1: " + timeA.dateFormat("M/d/y h:m:s") + "<br>");
        document.write("3 days, 6 hours and 10 minutes later:<br>");
        document.write("2: " + timeB.dateFormat("M/d/y h:m:s"));
    </script></p>
</body></html>
```

Listing 6.40 time_compute.htm File, Part 4 of 4

First, the program generates a single time entry, and then, it creates the second Date object using the dateAdd() method. This is another self-written extension of the Date object. Internally, a number of predefined methods are used (Section 6.4.5).

When you initiate the call, you can transfer a time value with a total of six partial values that will be added or subtracted. The six partial values represent the year, month, day, hour, minute, and second in sequence.

You can see this result in Figure 6.34 as well.

6.4.5 Second Extension of the Date Object

The custom `dateAdd()` method is used to extend the `Date` object and provides the option of adding or subtracting a time value from an existing time.

First, here's the code:

```
Date.prototype.dateAdd = function(year, month, day, hour, minute, second)
{
    const time = new Date();
    time.setFullYear(this.getFullYear() + year);
    time.setMonth(this.getMonth() + month);
    time.setDate(this.getDate() + day);
    time.setHours(this.getHours() + hour);
    time.setMinutes(this.getMinutes() + minute);
    time.setSeconds(this.getSeconds() + second);
    return time;
}
```

Listing 6.41 date.js File, Extension

The program assigns an anonymous function to the `dateAdd` property of the `Date.prototype` object. This makes `dateAdd()` a method, and six partial values of a time value are transferred as parameters. You access the current `Date` object for which the method is called via `this`.

A new `Date` object (`time`) is also created, and a reference to this object is returned as the result of the calculation. A total of six of the predefined `set` methods are called to calculate the individual partial values for the new object, and to calculate each partial value, the corresponding parameter is added to the corresponding partial value of the current object. If the partial values are too large or too small, they'll be converted accordingly. This applies, for example, to an entry such as −10 minutes or +50 days.

> **Note**
>
> You can find the `u_time_estimate` exercise in bonus chapter 1, section 1.17, in the downloadable materials for this book at *www.rheinwerk-computing.com/5875*.

6.5   Time-Controlled Processes

You've already become familiar with the alert(), prompt(), and confirm() methods of the window object. There are also the following other methods of the window object that enable you to control time processes, as required for games or simulations, for example:

- The setTimeout() method, which ensures that statements are called *once* after a certain time

- The setInterval() method, which makes sure the scheduled *repeated* call of statements after a certain time gets carried out

6.5.1   Starting Time-Controlled Processes

In the following program, the setTimeout() and setInterval() methods are used to display a current time specification with date and time:

```
... <head> ...
    <script src="number.js"></script>
    <script src="date.js"></script>
    <script>
        function intervalDateTime()
        {
            const d = new Date();
            document.getElementById("idInterval").firstChild.nodeValue =
                "With Interval: " + d.dateFormat("M/d/y h:m:s");
        }

        function timeoutDateTime()
        {
            const d = new Date();
            document.getElementById("idTimeout").firstChild.nodeValue =
                "With Timeout: " + d.dateFormat("M/d/y h:m:s");
            setTimeout(timeoutDateTime, 100);
        }
    </script>
</head>
<body>
    <p id="idInterval">With Interval:</p>
    <p id="idTimeout">With Timeout:</p>
    <script>
        setInterval(intervalDateTime, 100);
        setTimeout(timeoutDateTime, 100);
```

```
    </script>
</body></html>
```

Listing 6.42 process_digital_clock.htm File

The `setInterval()` method is called in the lower `script` container. The parameters of the method are a reference to a function and a duration in milliseconds.

The `intervalDateTime()` function is run repeatedly every 100 milliseconds from the call of the `setInterval()` method onward. A `Date` object with the current time is created in the function, and this time specification is formatted in the first paragraph of the document.

In addition, the `setTimeout()` method is called in the lower `script` container, with the same parameters.

The `timeoutDateTime()` function is run through *once* after 100 milliseconds by this call of the `setTimeout()` method, and a current time is also generated and formatted in the function. The `setTimeout()` method is called again, and in this way, the method calls itself every 100 milliseconds. Without this new call, the process would end at this point.

Figure 6.35 shows an output of this program.

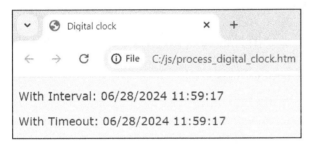

Figure 6.35 setInterval() and setTimeout() Methods

6.5.2 Starting and Ending Time-Controlled Processes

The `clearInterval()` method of the `window` object is used to end a time-controlled process that was started using the `setInterval()` method. In the following program, a countdown of 10 seconds runs down and stops at 0.

Here's the program:

```
... <head> ...
    <script>
        function countdown()
        {
            const timeB = new Date();
            const difference = (timeB - timeA) / 1000;
            const countdownNumber = 10 - difference;
```

```
        document.getElementById("idCountdown").firstChild.nodeValue
          = "Countdown: " + countdownNumber.toFixed(1);

        if(countdownNumber < 0.06)
        {
          clearInterval(intervalReference);
          document.getElementById("idCountdown")
            .firstChild.nodeValue = "Countdown: End";
        }
      }
    }
  </script>
</head>
<body>
  <p id="idCountdown">Countdown:</p>
  <script>
    const timeA = new Date();
    const intervalReference = setInterval(countdown, 100);
  </script>
</body></html>
```

Listing 6.43 process_countdown.htm File

First, the time at which the process starts is saved in the timeA variable, and then the setInterval() method is called. The latter also has a return value that references the process, and this reference is saved in a variable.

The countdown() function determines the time difference from the start in seconds, and it then calculates and outputs the remaining time. If this is less than 0.06 seconds, the clearInterval() method gets called to end the process. The saved reference of the setInterval() method is used as a parameter.

The output of the program is shown in Figure 6.36.

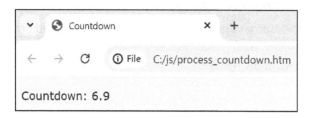

Figure 6.36 Countdown, clearInterval() Method

Note

You could also start multiple independent processes, and thanks to the individual references, each process can be controlled individually.

6.5.3   Controlling Processes

A time-controlled process is often started by an event, such as the clicking of a button. In most cases, you don't want the process to be able to be restarted during its active runtime, but only after it has ended. You would also like to have the option of pausing a process and then resuming it.

The following program has three buttons:

- When the **Start** button is clicked, the process starts, and any other clicks on it are ignored during the active runtime.

- Once the **Pause** button is clicked, the process halts, and it resumes running when the **Start** button is clicked.

- A click on the **Stop** button ends the process, regardless of whether it's currently active or paused.

Here's the program:

```
... <head> ...
  <script>
     let intervalReference, intervalActive = false;

     function output()
     {
        document.getElementById("idOutput").firstChild.nodeValue += "x";
     }

     function start()
     {
        if(intervalActive)
           return;
        intervalActive = true;
        intervalReference = setInterval(output, 500);
     }

     function pause()
     {
        intervalActive = false;
        clearInterval(intervalReference);
     }

     function stop()
     {
        intervalActive = false;
        clearInterval(intervalReference);
        document.getElementById("idOutput").firstChild.nodeValue = "Output: ";
```

```
        }
    </script>
</head>
<body>
    <p><input id="idStart" type="button" value="Start">
        <input id="idPause" type="button" value="Pause">
        <input id="idStop" type="button" value="Stop">
        <span id="idOutput">Output: </span></p>
    <script>
        document.getElementById("idStart").addEventListener("click", start);
        document.getElementById("idPause").addEventListener("click", pause);
        document.getElementById("idStop").addEventListener("click", stop);
    </script>
</body></html>
```

Listing 6.44 process_control.htm File

The intervalReference variable contains the reference to the process that can be terminated via this reference, and the intervalActive variable contains the information as to whether the process is active or not.

During the active runtime, the x character is appended to the existing output every 0.5 seconds. The start() function would be aborted immediately if activated during the active runtime.

The existing output remains in the pause() function. In the stop() function, on the other hand, it gets reset to the initial state.

Figure 6.37 shows an output of this program.

Figure 6.37 Controlling Processes

6.5.4 Slideshow and Single Image

With the help of time-controlled processes and switching between different images, you can display images both in single image mode and in a slideshow, as shown in Figure 6.38.

Figure 6.38 Slideshow and Single Image

You can click the **<<**, **<**, **>**, or **>>** button to immediately display the next image, the previous image, the first image, or the last image (respectively). The number of the current image and the total number of images are displayed. You can also start and stop a slideshow using the **Start slideshow** and **End slideshow** buttons.

Let's first look at the lower part of the program:

```
...
<body>
   <p><img id="idImage" src="im_paradise.jpg" alt="Paradise"></p>
   <p><input id="idFirst" type="button" value="&lt;&lt;">
      <input id="idBack" type="button" value="&lt;">
      <input id="idNext" type="button" value="&gt;">
      <input id="idLast" type="button" value="&gt;&gt;">
   <script>
      document.write("<span id='idOutput'>1/" + image.length + "</span></p>");
   </script>
   <p><input id="idStart" type="button" value="Start slideshow">
   <input id="idStop" type="button" value="End slideshow"></p>

   <script>
      document.getElementById("idFirst").addEventListener
         ("click", function() { move(0); } );
      document.getElementById("idBack").addEventListener
         ("click", function() { move(imageIndex - 1); } );
      document.getElementById("idNext").addEventListener
         ("click", function() { move(imageIndex + 1); } );
      document.getElementById("idLast").addEventListener
```

```
      ("click",function(){move(image.length - 1);});});
    document.getElementById("idStart").addEventListener("click", start );
    document.getElementById("idStop").addEventListener("click", stop );
  </script>
</body></html>
```

Listing 6.45 process_slideshow.htm File, Lower Part

The first four buttons use an anonymous function to call the move() function. A different parameter is passed each time the move() function is called. The image variable references an array that contains the names of the image files, and the imageIndex variable contains the index of the currently displayed image within the image array.

To display the next image, imageIndex gets increased by 1, and to display the preceding image, it gets decreased by 1. The first image has the index 0, and the last image has the index image.length - 1.

The next two buttons call the start() and stop() functions to start and stop the slideshow.

Now, here's the first part of the program:

```
... <head> ...
  <script>
    const image = ["im_paradise.jpg", "im_solar_eclipse.jpg",
        "im_wave.jpg", "im_winter.jpg", "im_tree.jpg"];
    let imageIndex = 0, timeoutReference, timeoutActive = false;

    function move(number)
    {
      imageIndex = number;
      if(imageIndex >= image.length)
        imageIndex = 0;
      else if(imageIndex < 0)
        imageIndex = image.length - 1;

      document.getElementById("idImage").src = image[imageIndex];
      document.getElementById("idOutput").firstChild.nodeValue =
        (imageIndex+1) + "/" + image.length;

      if(timeoutActive)
        timeoutReference = setTimeout(function() {move(imageIndex+1);}, 1000);
    }

    function start()
    {
```

```
      if(timeoutActive)
         return;
      timeoutActive = true;
      timeoutReference = setTimeout(function() {move(imageIndex+1);}, 500);
   }

   function stop()
   {
      clearTimeout(timeoutReference);
      timeoutActive = false;
   }
  </script>
</head>
. . .
```

Listing 6.46 process_slideshow.htm File, Upper Part

First, the program creates the array with the names of all image files. If you want to display your own images in a slideshow, you only need to change this part of the program.

The current index for this array is set to 0, and the timeoutReference variable references the time-controlled process. The timeoutActive variable stands for the information: the slideshow is currently active, or the slideshow is not running.

In the move() function, the value of imageIndex is set depending on the parameter of the function. If imageIndex gets too large, it will start again at 0, and if imageIndex gets too small, the system will start again at the end of the array.

The src property is given a new value. This changes the images, and the output gets changed accordingly.

If the slideshow is currently active, the move() function calls itself again after a short time using an anonymous function. The parameter is then determined using the current value of imageIndex.

If the slideshow is already running, the start() function gets exited immediately. Otherwise, the first image change for the slideshow is caused here. The slideshow is ended in the stop() function.

6.6   Other Data Structures

Data structures are used to store a group of thematically related data under a common name. The data consists of individual elements that usually have the same type but can also have different types.

In Section 6.1, you learned about the *array* data structure. In this section, we add more data structures called sets and maps.

6.6.1   Sets

A *set* is a collection of unique values, which means that each value only occurs once within a set. In JavaScript, the Set object is used to create a set. A set has properties and methods that you can use to access its values.

The following program shows some typical operations with sets:

```
...
<body><p>
   <script>
      const s = new Set([23, 5, 28, -4, 5, 8, 23]);
      s.add(5);
      s.add(36);

      document.write("Values with forEach(): ");
      function output(value)
      {
         document.write(value + " ");
      }
      s.forEach(output);
      document.write("<br>");

      let v = 23;
      if(s.has(v))
         document.write("Value " + v + " is in the set<br>");
      v=24;
      if(!s.has(v))
         document.write("Value " + v + " is not in the set<br>");

      document.write("Values with values(): ");
      const it = s.values();
      for(let i=0; i<s.size; i++)
         document.write(it.next().value + " ");
      document.write("<br>");

      document.write("Number of values: " + s.size + "<br>");
      s.delete(8);
      document.write("After deleting a value: " + s.size + "<br>");
      s.clear();
      document.write("After clearing the set: " + s.size);
   </script></p>
</body></html>
```

Listing 6.47 set.htm File

You can fill a set with values when it gets created. For this purpose, an enumeration is passed in square brackets containing the possible values. If one of the values occurs twice, it's only added to the set once.

The add() method is used to add individual values. Here, the same principle applies: if the value already exists, it will not be added.

You can use the forEach() method to run through a set completely, in the order in which the values are added. The reference to a function that is executed for each value serves as a parameter. Here, it's the output() method for displaying the value. What is not visible to us is that the forEach() method makes sure the value gets passed to the output() method as the first parameter.

The has() method determines whether a certain value is in the set or not.

The values() method returns an enumeration object of the *iterator* type, which we refer to as an *iterator* for short. The next() method can be called for objects of this type, and it provides a reference to the next element in the enumeration. The value of this element can be determined using the value property. The value of the size property corresponds to the number of values in a set, and this can be used in a for loop to output all values of the set.

The delete() method is used to delete a specific value from the set, and the clear() method deletes all values from the set. The output of the program is shown in Figure 6.39.

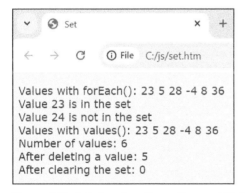

Figure 6.39 Set: Creation, Properties, and Methods

6.6.2 Maps

In JavaScript, the Map object is used to create an associative array, which we refer to as a *map* for short. A map consists of a collection of key-value pairs, and you can access the individual values by using a key that consists of a string, for example. In comparison, with a standard array (Section 6.1), you access the individual values using a numerical index. A map has properties and methods with which you can access its keys and values, and there are some parallels to the properties and methods of a set.

The following program shows some typical operations that involve maps:

```
...
<body><p>
  <script>
      const m = new Map([["Peter", 31], ["Julia", 28], ["Warren", 35]]);
      m.set("Sandy", 38);
      m.set("Warren", 34);

      document.write("Elements with forEach(): ");
      function output(value, key)
      {
          document.write(key + ":" + value + " ");
      }
      m.forEach(output);
      document.write("<br>");

      let s = "Julia";
      if(m.has(s))
          document.write("Key:" + s + ", Value:" + m.get(s) + "<br>");
      s = "Ben";
      if(!m.has(s))
          document.write("Key:" + s + ", not available" + "<br>");

      document.write("Values with values(): ");
      const v = m.values();
      for(let i=0; i<m.size; i++)
          document.write(v.next().value + " ");
      document.write("<br>");

      document.write("Keys with keys(): ");
      const k = m.keys();
      for(let i=0; i<m.size; i++)
          document.write(k.next().value + " ");
      document.write("<br>");

      document.write("key-value pairs with entries(): ");
      const e = m.entries();
      for(let i=0; i<m.size; i++)
          document.write(e.next().value + " ");
      document.write("<br>");

      document.write("Number of elements: " + m.size + "<br>");
      m.delete("Warren");
```

```
        document.write("After deleting an element: " + m.size + "<br>");
        m.clear();
        document.write("After clearing the map: " + m.size);
    </script></p>
</body></html>
```

Listing 6.48 map.htm File

You can fill a map when you create it. For this purpose, an enumeration of key-value pairs is transferred. Each pair and the list as a whole are enclosed in square brackets.

The set() method is used to add individual pairs. If the key for the pair already exists, the corresponding value will be overwritten.

You can use the forEach() method to run through a map completely in the order in which the pairs are added. The reference to a function that is executed for each value serves as a parameter, and here, it's the output() method for displaying the pairs. What is not visible to us is that the forEach() method makes sure that the key and value are passed as parameters to the output() method.

The has() method determines whether a specific key is in the map or not, and the get() method can be used to access the value of a specific key.

The values() method returns an iterator (Section 6.6.1) with all the values of the map, and an iterator with all the keys of the map is provided by the keys() method. The entries() method returns an iterator with all pairs of the map, and the value of the size property corresponds to the number of pairs in a map. In a for loop, these relationships can be used to output all values, keys, or pairs of the map. The delete() method is used to delete a specific pair from the map, and the clear() method deletes all pairs from the map.

The output of the program is shown in Figure 6.40.

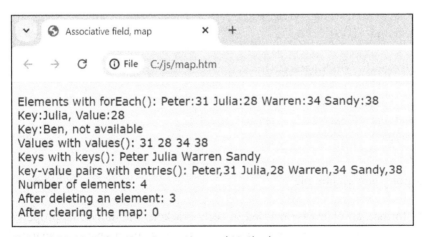

Figure 6.40 Map: Creation, Properties, and Methods

6.7   JavaScript Object Notation

JavaScript Object Notation (JSON) is the name for a compact notation that can be applied to objects and arrays. We refer to this notation as *JSON format.*

Modern browsers are familiar with the *JSON object*, which has static methods that you can use to convert an object that is in JSON format to a transportable string. In turn, you can create an object in JSON format from such a string.

A transportable string can be processed not only by JavaScript but also by other languages, which simplifies the transport of data between different applications.

6.7.1   JSON Format and JSON Objects

In the following example, the program creates an object in JSON format and then converts it using the JSON object.

Here's the program:

```
...
<body><p>
   <script>
      const dodge = { "color":"Red", "speed":50.2 };
      document.write(dodge.color + " " + dodge.speed + "<br>");

      const tx = JSON.stringify(dodge);
      document.write(tx + "<br>");

      const renault = JSON.parse(tx);
      document.write(renault.color + " " + renault.speed + "<br>");
      renault.speed += 2.5;
      document.write("Faster: " + renault.speed);
   </script></p>
</body></html>
```

Listing 6.49 json_compact.htm File

In the JSON format, an object is enclosed in curly brackets, the individual property-value pairs are separated by commas, the property and the value of a pair are separated by a colon, and the name of the property is placed in double quotation marks. This remains the same if the value is a string, but if it's a numerical value, it is noted without quotation marks. Decimal places are separated using a decimal point.

In the program, the dodge object, which is in JSON format, is converted into the transportable tx string using the stringify() method of the JSON object. The reverse process follows: the tx string is converted into the renault object using the parse() method of the JSON object, which is then available in JSON format.

The value of the speed property is recognized as a numerical value and can therefore also be included in a calculation.

The output of the program looks as shown in Figure 6.41.

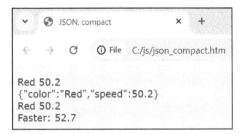

Figure 6.41 JSON Format and JSON Object

Note

ECMAScript 2024 introduces the option of making objects in JSON format immutable using the # character. For example, const dodge = #{ "color": "Red" }; means that the property of the object can no longer be changed, and the dodge.color = "Blue"; statement then results in an error.

6.7.2 Custom Classes and JSON Methods

Thanks to an intermediate step, the conversion methods of the JSON object are also available for the Car class.

Here's the program:

```
... <head> ...
   <script>
      class Car
      {
         constructor(c, s)
         {
            this.color = c;
            this.speed = s;
         }
      }
```

```
        toString()
        {
            return this.color + " " + this.speed;
        }
    }
    </script>
</head>
<body><p>
    <script>
        const dodge = new Car("Red", 50.2);
        document.write(dodge + "<br>");

        const tx = JSON.stringify(dodge);
        document.write(tx + "<br>");

        const x = JSON.parse(tx);
        const renault = new Car(x.color, x.speed);
        document.write(renault + "<br>");
        renault.speed += 2.5;
        document.write("Faster: " + renault.speed);
    </script></p>
</body></html>
```

Listing 6.50 json_class.htm File

The program converts the object of the custom Car class into a transportable string using the stringify() method of the JSON class.

As an intermediate step, an object (x) in JSON format is created from the transportable string using the parse() method. A new object of the Car class is then created using the properties and values of object x.

The output of the program also looks as shown in Figure 6.41.

6.7.3   JSON Object and Arrays

The JSON object also allows you to use arrays, as the following program shows:

```
...
<body><p>
    <script>
        const city = ["Berlin", "Hamburg", "Munich"];
        document.write(city + "<br>");

        const tx = JSON.stringify(city);
        document.write(tx + "<br>");
```

```
      const cityNew = JSON.parse(tx);
      document.write(cityNew);
   </script></p>
</body></html>
```

Listing 6.51 json_array.htm File

The program converts an array of strings (`city`) into the transportable `tx` string using the `stringify()` method of the JSON object. Then, it creates an array of strings (`cityNew`) using the `parse()` method of the JSON object.

The output of the program is shown in Figure 6.42.

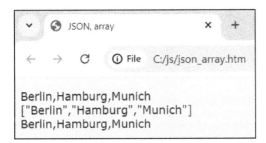

Figure 6.42 JSON Object and Arrays

6.7.4   JSON Format and Arrays

The following program uses an array of objects that is available in JSON format:

```
...
<body><p>
   <script>
      const cr = [{"color":"Red", "speed":50.2},
                  {"color":"Blue", "speed":85.3},
                  {"color":"Yellow", "speed":65.1}];
      for(let i=0; i<cr.length; i++)
         document.write(cr[i].color + " " + cr[i].speed + " # ");
      document.write("<br>");

      const tx = JSON.stringify(cr);
      document.write(tx + "<br>");

      const crNew = JSON.parse(tx);
      for(let i=0; i<crNew.length; i++)
         document.write(crNew[i].color + " " + crNew[i].speed + " # ");

      document.write("<br>Faster: ");
      for(let i=0; i<crNew.length; i++)
```

```
        {
            crNew[i].speed += 2.5;
            document.write(crNew[i].speed + " # ");
        }
    </script></p>
</body></html>
```

Listing 6.52 json_array_compact.htm File

The entire array of objects in JSON format (cr) is enclosed in rectangular brackets, and the individual objects are enclosed in curly brackets.

The transportable tx string is created using the stringify() method of the JSON object, and the tx string is converted into a new array of objects in JSON format using the parse() method of the JSON object.

Here, too, the value of the speed property is recognized as a numerical value and can therefore be included in a calculation.

The output of this part of the program is shown in Figure 6.43.

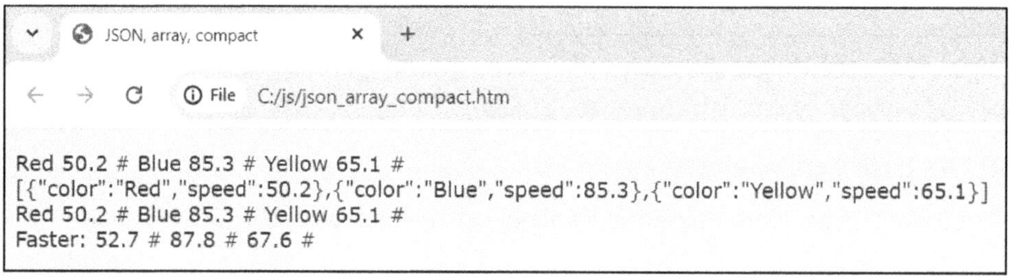

Figure 6.43 JSON Format and Arrays

6.7.5   Custom Classes and Arrays

The following program uses an array of objects of the Car class. The conversion methods of the JSON object are also available here, thanks to an intermediate step for the Car class.

Here's the program:

```
...  <head> ...
    <script>
        class Car
        {
            constructor(c, s)
            {
                this.color = c;
                this.speed = s;
            }
```

```
            toString()
            {
                return this.color + " " + this.speed;
            }
        }
    </script>
</head>
<body><p>
    <script>
        const cr = [new Car("Red", 50.2),
            new Car("Blue", 85.3), new Car("Yellow", 65.1)];
        for(let i=0; i<cr.length; i++)
            document.write(cr[i] + " # ");
        document.write("<br>");

        const tx = JSON.stringify(cr);
        document.write(tx + "<br>");

        const crNew = new Array(3);
        const x = JSON.parse(tx);
        for(let i=0; i<crNew.length; i++)
        {
            crNew[i] = new Car(x[i].color, x[i].speed);
            document.write(crNew[i] + " # ");
        }

        document.write("<br>Faster: ");
        for(let i=0; i<crNew.length; i++)
        {
            crNew[i].speed += 2.5;
            document.write(crNew[i].speed + " # ");
        }
    </script></p>
</body></html>
```

Listing 6.53 json_array_class.htm File

The program converts the array of objects of the custom Car class into a transportable string using the stringify() method of the JSON class.

Then, an empty array (crNew) is created in the required size.

As an intermediate step, an array of objects (x) in JSON format is generated from the transportable string. The new objects of the Car class are then created using the properties and values of the elements of the x array and saved in the crNew array.

6

The output of the program also looks like the one shown in Figure 6.43. JSON files are imported in Chapter 7, Section 7.4.

Note

In bonus chapter 3 in the downloadable materials for this book (found at *www.rhein-werk-computing.com/5875*), you can find some sample projects that contain elements from throughout Chapter 6.

Chapter 7
Changes Using Ajax

Ajax technology makes it possible to exchange parts of a website with data from the web server without having to completely rebuild the displayed document.

Asynchronous JavaScript and XML (Ajax) provides asynchronous data transfer between a browser and a web server. It allows you to change parts of a website without having to create and submit the entire document again, which reduces development costs and network traffic and makes websites faster and more flexible. Ajax technology is a standard component of many websites.

The process looks like this: A request is sent to the web server from a website as a result of an event, and the web server returns a response. The response is analyzed in the document that is still displayed, and the analysis leads to a part of the document being changed. The central component of the entire process is an object of the `XMLHttpRequest` type.

All documents in this chapter are accessed via a web server from which you request data to insert into the pages. In this chapter, I retrieve the documents via my domain *theisweb.de* from the *jse* directory. The data can be available on the web server in various forms:

- As a PHP program (Section 7.1)
- As text in a text file (as shown at the end of Section 7.1)
- As an XML file (Section 7.3)
- As a JSON object in a text file (Section 7.4)

7.1 Hello Ajax

Using the first example, `Hello Ajax`, I will explain the basic process that applies to the entire chapter.

In the example, an HTML document containing a hyperlink is displayed (see Figure 7.1). Here, the hyperlink doesn't reference another document, so the hyperlink target consists only of the # character. Instead, the hyperlink is used to trigger an event.

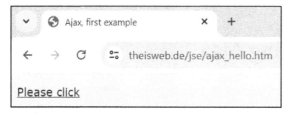

Figure 7.1 Permanently Displayed Document

The document contains a further paragraph. Clicking on the hyperlink sends a request to the web server, which sends a text in response, and the text becomes the content of the paragraph and appears in the document as you see it in Figure 7.2. The rest of the document isn't changed.

Figure 7.2 Document Supplemented from PHP File

Let's first look at the HTML document with the JavaScript code:

```
... <head> ...
   <script>
      function request()
      {
         const req = new XMLHttpRequest();
         req.open("get", "ajax_hello.php", true);
         // req.open("get", "ajax_hello.txt", true);
         req.onreadystatechange = evaluate;
         req.send();
      }

      function evaluate(e)
      {
         if(e.target.readyState == 4 && e.target.status == 200)
            document.getElementById("idParagraph")
               .firstChild.nodeValue = e.target.responseText;
      }
   </script>
</head>
```

```
<body>
    <p><a id="idLink" href="#">Please click</a></p>
    <p id="idParagraph"> </p>
    <script>
        document.getElementById("idLink").addEventListener("click", request);
    </script>
</body></html>
```

Listing 7.1 ajax_hello.htm File

In addition, here's the responding PHP program:

```
<?php
    header("Content-type: text/html; charset=utf-8");
    echo "PHP file: Hello Ajax";
?>
```

Listing 7.2 ajax_hello.php File

The *ajax_hello.htm* file contains the hyperlink and the paragraph, and clicking on the hyperlink calls the request() function. First, a new XMLHttpRequest object is created.

The open() method of this object opens communication with another file on the web server, in this case, with *ajax_hello.php* and the GET method. The third parameter of the open() method is usually set to true. This ensures that communication is handled asynchronously, which means other processes don't have to wait for the end of the request.

The onreadystatechange event handler is assigned a reference to a named or anonymous function, in this case, to the evaluate() function. After sending, the XMLHttpRequest object changes its status several times, and the event handler responds to this change event. In a nutshell, the evaluate() method is called each time the status changes.

The send() method of the XMLHttpRequest object sends the request to the web server. No other data will be transmitted initially, and the transfer should only take place after the event handler has been registered. In this way, no event can go unnoticed.

An event object is transferred to the evaluate() method, and the target property of this object references the XMLHttpRequest object. First, its readystate and status properties are taken into consideration. The evaluation only makes sense after readystate has assumed the value 4 and status has assumed the value 200. The status property represents the value of the status code of the *Hypertext Transfer Protocol* (HTTP). The 200 code stands for *OK*, 404 for *Page not found*, 500 for *Internal Server Error*, and so on.

The responseText property of the XMLHttpRequest object contains the response from the web server—in this case, the Hello Ajax text, which is output using the PHP keyword, echo. This text is set as the new content of the paragraph.

You can also request the content of a text file from the web server and insert it into an existing document. To do this, you must swap the call of the open() method in the request() function (i.e., set the comment characters for the second line and remove them for the third line). The text file looks as follows:

```
TXT file: Hello Ajax
```

Listing 7.3 ajax_hello.txt File

> **Note**
>
> The following applies to this transmission between browser and web server:
>
> - Communication is regulated using HTTP. HTTPS is the secure, encrypted further development of HTTP.
> - The *request* from the browser to the web server asks for data, while the *response* is the answer from the web server to the browser.
> - There are several methods for a request. The GET method transfers smaller amounts of data (up to 255 bytes), while the POST method transfers larger amounts of data.

7.2   Sending Parameters

This section contains an example in which parameters are sent to the web server. The response of the PHP program depends on the data in these parameters.

An HTML document with two hyperlinks and an empty paragraph is displayed (see Figure 7.3).

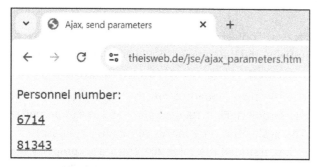

Figure 7.3 Call with Parameters

When you click the first hyperlink, the system requests data on the person with personnel number 6714, and when you click the second hyperlink, it requests the data for the person with personnel number 81343. The data on the person then appears below the hyperlinks (see Figure 7.4 and Figure 7.5).

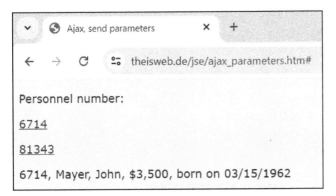

Figure 7.4 All Data for Personnel Number 6714

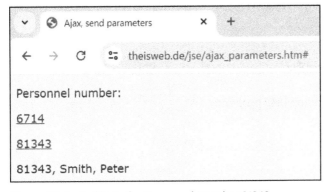

Figure 7.5 Part of Data for Personnel Number 81343

Let's first look at the HTML document with the JavaScript code:

```
... <head> ...
   <script>
      function request(personnel_number, scope)
      {
         const req = new XMLHttpRequest();
         req.open("get", "ajax_parameters.php?pno=" + personnel_number
            + "&scope=" + scope, true);
         req.setRequestHeader("Content-Type",
            "application/x-www-form-urlencoded");
         req.onreadystatechange = evaluate;
         req.send();
      }

      function evaluate(e)
      {
         if(e.target.readyState == 4 && e.target.status == 200)
            document.getElementById("idParagraph").firstChild.nodeValue =
```

```
                    e.target.responseText;
        }
    </script>
</head>
<body>
    <p>Personnel number:</p>
    <p><a id="idLink0" href="#">6714</a></p>
    <p><a id="idLink1" href="#">81343</a></p>
    <p id="idParagraph"> </p>
    <script>
        document.getElementById("idLink0").addEventListener
            ("click", function() { request(6714, "all"); });
        document.getElementById("idLink1").addEventListener
            ("click", function() { request(81343, "part"); });
    </script>
</body></html>
```

Listing 7.4 ajax_parameters.htm File

In addition, here's the responding PHP program:

```
<?php
    header("Content-type: text/html; charset=utf-8");

    if($_GET["pno"] == 6714)
    {
        if($_GET["scope"] == "all")
            echo "6714, Mayer, John, $3,500, born on 03/15/1962";
        else
            echo "6714, Mayer, John";
    }
    else if($_GET["pno"] == 81343)
    {
        if($_GET["scope"] == "all")
            echo "81343, Smith, Peter, $3,750, born on 04/12/1958";
        else
            echo "81343, Smith, Peter";
    }
?>
```

Listing 7.5 ajax_parameters.php File

Clicking on one of the two hyperlinks calls the request() function, which transmits two parameters: a number and a string.

You use the setRequestHeader() method of the XMLHttpRequest object to define the form in which this data is transmitted. In this case, it gets appended to the address of the website. The parameter of the open() method of the XMLHttpRequest object corresponds to the contents of a so-called search string, and the individual parts of the search string are separated from each other with the & sign. In the case of the first hyperlink, the search string is pno=6714&scope=all.

After the data has been sent, the $_GET["pno"] and $_GET["scope"] elements are available in the PHP file with their respective values. Within a branch, a decision is made as to which response is sent back.

7.3 Reading an XML File

Extensible Markup Language (XML) is a widespread, platform-independent data format that is suitable for universal data exchange. You can use Ajax to incorporate into your documents the content of XML files that are generated with external programs and stored on the web server.

XML is actually a specification of how a language must be structured. There's a hierarchy of nodes with child nodes in XML, as you've already seen for HTML in Chapter 5. The elements of an XML file are also created using tags, containers, and attributes. XML files can be edited in a simple text editor.

7.3.1 Single Object

You can use Ajax to integrate the content of an XML file on the web server into a document.

Here's the XML file first:

```
<?xml version="1.0" encoding="UTF-8"?>
<car>
   <color>Red</color>
   <speed>50</speed>
   <power displacement="1600" cylinders="4">76.2</power>
</car>
```

Listing 7.6 ajax_xml_single.xml File

Modern browsers can display the structure of an XML file directly, and this holds true regardless of whether the XML file is retrieved via a web server or directly from the local hard disk. In Figure 7.6, you can see the specified XML file.

An XML file can be linked to another file for formatting the XML data, but this is not the case in the examples in this chapter. For this reason, the note with the text This XML file does not appear … is provided in Figure 7.6.

269

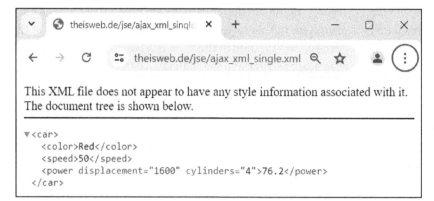

Figure 7.6 Structure of XML File

The first line of the XML code identifies the content as XML. The following character encoding specifies the character set used, as for HTML documents. There's a main element or root node in every XML file, and you can freely choose the names of the individual nodes yourself, as long as you adhere to the XML rules and remain true to your own structure.

The root node here is the car element. A car has three properties with values that are defined as child nodes, and the third child node has two attributes, each of which has values. If a value is a number with decimal places, these must be separated by a decimal point.

This data is read from the XML file after you click on the hyperlink using Ajax, and it fills the paragraph (see Figure 7.7).

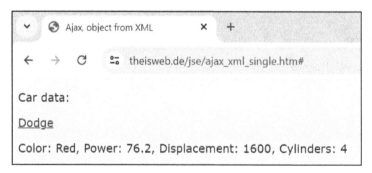

Figure 7.7 Data of Single Object

Here's the corresponding program:

```
... <head> ...
  <script>
    function request()
    {
      const req = new XMLHttpRequest();
```

```
        req.open("get", "ajax_xml_single.xml", true);
        req.onreadystatechange = evaluate;
        req.send();
    }

    function evaluate(e)
    {
        if(e.target.readyState == 4 && e.target.status == 200)
        {
            const response = e.target.responseXML;
            const kcolor = response.getElementsByTagName("color")[0];
            const kpower = response.getElementsByTagName("power")[0];
            document.getElementById("idData").firstChild.nodeValue =
                "Color: " + kcolor.firstChild.nodeValue
                    + ", Power: " + kpower.firstChild.nodeValue
                    + ", Displacement: " + kpower.getAttribute("displacement")
                    + ", Cylinders: " + kpower.getAttribute("cylinders");
        }
    }
    </script>
</head>
<body>
    <p>Car data:</p>
    <p><a id="idLink" href="#">Dodge</a></p>
    <p id="idData"> </p>
    <script>
        document.getElementById("idLink").addEventListener("click", request);
    </script>
</body></html>
```

Listing 7.7 ajax_xml_single.htm File

In the request() function, the open() method opens the communication with the responding XML file. An XML document is requested instead of a text document, which is why the evaluate() function uses the responseXML property and not the responseText property.

In the following paragraphs, you'll see many parallels to the contents of Chapter 5. The getElementsByTagName() method of the document object returns an array with references to all XML elements with the required tag, and in this case, it's done for the color tag and the power tag. In the example, there's only one array element with the index 0.

The first child node of the first element is output using the firstChild property of a node object. The value of two different attributes is then determined and output using the getAttribute() method of a node object.

7.3.2   Collection of Objects

In the following XML file, you can see a collection of objects that have a similar structure. Ajax is used to integrate the contents of this XML file into the HTML document.

Here's the XML file first:

```xml
<?xml version="1.0" encoding="UTF-8"?>
<collection>
   <car>
      <color>Red</color>
      <speed>50</speed>
      <power displacement="1600" cylinders="4">76.2</power>
   </car>
   <car>
      <color>Yellow</color>
      <speed>65</speed>
      <power displacement="1800" cylinders="4">85.0</power>
   </car>
</collection>
```

Listing 7.8 ajax_xml_collection.xml File

In Figure 7.8, you can see the XML file without formatting.

Figure 7.8 Structure of XML File

There are several nodes of the car type. Another node must be added as a parent so that there's still only one main element or root node. This element has the collection tag and two child nodes of the car type described earlier.

After you click on one of the hyperlinks in Figure 7.9, the initially empty paragraph is filled using Ajax.

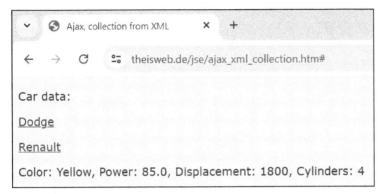

Figure 7.9 Data of Object from Collection

Here's the corresponding program:

```
...  <head> ...
   <script>
      function request(x)
      {
         const req = new XMLHttpRequest();
         req.open("get", "ajax_xml_collection.xml", true);
         req.onreadystatechange = function(e) { evaluate(e, x); };
         req.send();
      }

      function evaluate(e, x)
      {
         if(e.target.readyState == 4 && e.target.status == 200)
         {
            const response = e.target.responseXML;
            const kcolor = response.getElementsByTagName("color")[x];
            const kpower = response.getElementsByTagName("power")[x];
            document.getElementById("idData").firstChild.nodeValue =
               "Color: " + kcolor.firstChild.nodeValue
                  + ", Power: " + kpower.firstChild.nodeValue
                  + ", Displacement: " + kpower.getAttribute("displacement")
                  + ", Cylinders: " + kpower.getAttribute("cylinders");
         }
      }
   </script>
</head>
<body>
```

```
<p>Car data:</p>
<p><a id="idLink0" href="#">Dodge</a></p>
<p><a id="idLink1" href="#">Renault</a></p>
<p id="idData"> </p>
<script>
   document.getElementById("idLink0").addEventListener
      ("click", function() { request(0); } );
   document.getElementById("idLink1").addEventListener
      ("click", function() { request(1); } );
</script>
</body></html>
```

Listing 7.9 ajax_xml_collection.htm File

When you click one of the hyperlinks, the corresponding value is passed as a parameter to the `request()` function. This value then gets passed further on to the `evaluate()` function, using an anonymous function. This way, the data of the matching object can be determined from the collection and returned in order to fill the initially empty paragraph.

7.3.3   Search Suggestions

One of the typical uses of Ajax is to display suggestions when you enter a search string into an input field. I will demonstrate this in the following example.

An XML file contains a collection of data on countries in the European Union. When you enter characters into the search field of the program, the data on those countries whose first letters match the characters you've already entered is displayed as a search aid. Any change in the search field can lead directly to a change in the display.

In Figure 7.10, you can see the result after you enter "B" into the search field. The entry is extended to "Be" in Figure 7.11, and if you delete the "e," it will again look like Figure 7.10.

Figure 7.10 After Entering "B"

Figure 7.11 After Entering "Be"

Here's the content of the XML file first:

```
<?xml version="1.0" encoding="UTF-8"?>
<countries>
    <country>
        <name>Belgium</name>
        <area>30528</area>
        <city>Brussels</city>
    </country>
    <country>
        <name>Bulgaria</name>
        <area>110994</area>
        <city>Sofia</city>
    </country>
    <country>
        <name>Denmark</name>
        <area>43094</area>
        <city>Copenhagen</city>
    </country>
</countries>
```

Listing 7.10 ajax_eu.xml File

In Figure 7.12, you can see the XML file without any formatting.

The main element or the root node has the countries tag and child nodes of the country type. For each object of the country type, there are the name, area, and city child nodes.

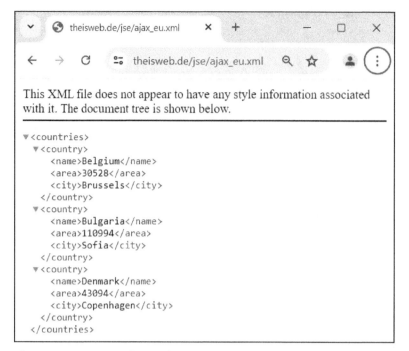

Figure 7.12 Structure of XML File

Here's the first part of the program:

```
...
<body>
    <p id="idParagraph"><input id="idInput"> Search for EU countries</p>
    <script>
        document.getElementById("idInput")
            .addEventListener("keydown", request);
    </script>
</body></html>
```

Listing 7.11 ajax_eu.htm File, Lower Part

The paragraph with the idParagraph ID contains two child nodes: the search field with the idInput ID and the Search for EU countries text. There's also an event handler for the keydown event that is triggered when you press a key.

Let's now take a look at the upper part of the program with the JavaScript functions:

```
...  <head> ...
    <script>
        function request()
        {
            const req = new XMLHttpRequest();
```

```
        req.open("get", "ajax_eu.xml", true);
        req.onreadystatechange = evaluate;
        req.send();
    }

    function evaluate(e)
    {
        if(e.target.readyState == 4 && e.target.status == 200)
        {
            const par = document.getElementById("idParagraph");
             while(par.childNodes.length > 2)
                par.removeChild(par.lastChild);

            if(document.getElementById("idInput").value == "")
                return;

            const response = e.target.responseXML;
            const nameAll = response.getElementsByTagName("name");
            const search = document.getElementById("idInput").value;

            for(let i=0; i<nameAll.length; i++)
            {
                const name = nameAll[i].firstChild.nodeValue;

                if(search == name.substr(0,search.length))
                {
                  const area = response.getElementsByTagName(
                      "area")[i].firstChild.nodeValue;
                  const city = response.getElementsByTagName(
                      "city")[i].firstChild.nodeValue;

                  const text = document.createTextNode(name + ", "
                      + area + " sq mi, " + city);
                  const divElement = document.createElement("div");
                  divElement.appendChild(text);
                  par.appendChild(divElement);
                }
            }
        }
    }
    </script>
</head>
...
```

Listing 7.12 ajax_eu.htm File, Upper Part

In the `request()` function, the `open()` method opens the communication with the responding XML file. The response is provided in the `responseXML` property.

The old search suggestions are removed in the `evaluate()` function. These were additional child nodes of the paragraph with the `idParagraph` ID. Using a `while` loop, these additional child nodes are deleted individually, starting with the last child node. The deletion process runs until only the two original child nodes remain: the search field with the `idInput` ID and the `Search for EU countries` text.

If the search field is completely empty, the `evaluate()` function is exited and no search help gets displayed. However, if the search field still contains characters, a field with all country names will be determined from the XML response. The elements of this field get checked, and if the characters you've entered so far match the first letters of a country name, the data for this country gets displayed.

First, for this purpose, the information on the area and capital of this country is determined. Next, a text node that summarizes the country data in a readable form is created, and this text node is assigned to a newly created HTML element node of the `div` type as a child node. Finally, this `div` node is added to the paragraph with the `idParagraph` ID as a new child node.

Note

Here are some tips for possible improvements:

- You could use `setTimeout()` to wait for a short time before making the request after you've clicked a button. In this way, multiple changes could be implemented in the request without calling the function each time.
- The content of the XML file could be generated dynamically using PHP and a database; more information would then be available for selection.

7.4   Reading a JSON File

You've already become familiar with JSON in Chapter 6, Section 6.7. The JSON format is an alternative to XML as a universal data exchange format, and here, I will show you how JSON simplifies the transport of data between different applications.

A text file contains data in JSON format, which is saved using PHP, for example. We'll integrate this data into an existing HTML document using Ajax.

7.4.1   Single Object

First, we want to look at a single object in JSON format in a text file. We'll integrate this object into a website using Ajax, and the text file looks like this:

```
{"color":"Red","speed":50.2}
```

Listing 7.13 ajax_json_single.txt File

The text file contains the entire object in JSON format. Its data is read from the text file using Ajax after the hyperlink has been clicked and fills the paragraph (see Figure 7.13).

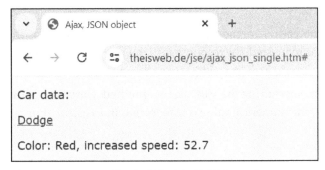

Figure 7.13 Data on Single Object in JSON Format

Here's the corresponding program:

```
... <head> ...
   <script>
      function request()
      {
         const req = new XMLHttpRequest();
         req.open("get", "ajax_json_single.txt", true);
         req.onreadystatechange = evaluate;
         req.send();
      }

      function evaluate(e)
      {
         if(e.target.readyState == 4 && e.target.status == 200)
         {
            const response = JSON.parse(e.target.responseText);
            document.getElementById("idOutput").firstChild.nodeValue
               = "Color: " + response.color
               + ", increased speed: " + (response.speed + 2.5);
         }
      }
   </script>
</head>
<body>
   <p>Car data:</p>
   <p><a id="idLink" href="#">Dodge</a></p>
```

```
    <p id="idOutput"> </p>
    <script>
        document.getElementById("idLink").addEventListener("click", request);
    </script>
</body></html>
```

Listing 7.14 ajax_json_single.htm File

In the `request()` function, the `open()` method opens the communication with the responding text file. As a text is requested, the response is in the `responseText` property in the `evaluate()` function.

The string is converted into an object using the `JSON.parse()` method. The values of the object's properties are output, the numerical value is recognized, and a calculation can be performed with it.

7.4.2   Collection of Objects

The following text file contains an array of objects in JSON format. Ajax is used to integrate the content into an HTML document:

```
[ {"color":"Red","speed":50.2},
  {"color":"Blue","speed":85.3},
  {"color":"Yellow","speed":65.1} ]
```

Listing 7.15 ajax_json_collection.txt File

After you click one of the hyperlinks in Figure 7.14, Ajax fills the text field below it.

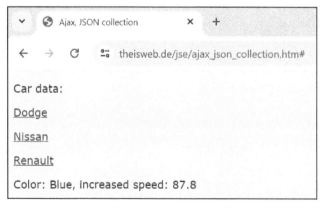

Figure 7.14 Data on Object in JSON Format from Collection

Here's the corresponding program:

```
... <head> ...
    <script>
```

```
    function request(x)
    {
        const req = new XMLHttpRequest();
        req.open("get", "ajax_json_collection.txt", true);
        req.onreadystatechange = function(e) { evaluate(e, x); };
        req.send();
    }

    function evaluate(e, x)
    {
        if(e.target.readyState == 4 && e.target.status == 200)
        {
            const response = JSON.parse(e.target.responseText);
            document.getElementById("idOutput").firstChild.nodeValue
                = "Color: " + response[x].color + ", increased speed: "
                + (response[x].speed + 2.5);
        }
    }
    </script>
</head>
<body>
    <p>Car data:</p>
    <p><a id="idLink0" href="#">Dodge</a></p>
    <p><a id="idLink1" href="#">Nissan</a></p>
    <p><a id="idLink2" href="#">Renault</a></p>
    <p id="idOutput"> </p>
    <script>
        for(let i=0; i<3; i++)
            document.getElementById("idLink" + i).addEventListener
                ("click", function() { request(i); } );
    </script>
</body></html>
```

Listing 7.16 ajax_json_collection.htm File

When you click one of the hyperlinks, the program passes the corresponding value as a parameter to the request() function. This value then gets passed further on to the eval-uate() function, using an anonymous function. In this way, the appropriate element can be determined from the array. The data of this element is returned and fills the paragraph, after a calculation if necessary.

Chapter 8
Design Using Cascading Style Sheets

*The elements of websites are formatted and positioned in a standard-
ized way using CSS. JavaScript allows for changes and effects, including
animation.*

Cascading Style Sheets (CSS) are formatting templates that complement each other.
Often only the term *style sheets* is used, without *cascading*.

Style sheets give us the opportunity to do the following:

- Separate formatting and content of an HTML document.
- Define central, standardized formatting.
- Supplement or overwrite these with local formatting.
- Implement formatting that goes beyond HTML.
- Position elements anywhere in the document.

Extensive websites with many individual webpages normally use centralized formatting. They give the website a uniform look, save code, and are easy to change for the
entire website. In this book, I will only explain a small part of the great variety of CSS,
which is constantly being expanded.

You can change style sheets dynamically using JavaScript. This opens up interesting
possibilities, particularly in the area of positioning elements such as animations.

8.1   Structure and Rules

This section describes the structure of style sheets, the combination of different style
sheets, and some rules for using style sheets.

The three files in this section do not include the *js5.css* file, which ensures a uniform display using style sheets in most of the examples in this book. Instead, this section uses
its own style sheets to demonstrate the effects.

8.1.1   Locations and Selectors

You can define style sheets in the following different locations:

- *External*: In an external file that can be integrated and leads to uniform formatting in many documents
- *Embedded*: In the header of a document using a `style` container so that the formatting is uniform throughout the document
- *Inline*: In an HTML tag using the `style` attribute so that the formatting only applies to this position in the document

You can also assign style sheets to specific parts of a document by using selectors in the following ways:

- HTML selectors assign the style sheet to all elements with the same HTML tag.
- Class selectors assign the style sheet to all HTML tags to which the relevant class is assigned.
- ID selectors assign the style sheet to individual elements, each of which has a unique ID.

We explain the various locations and selectors in the following paragraphs, based on an example. Style sheets can consist of multiple CSS properties, but in this example, we initially use only one CSS property: `text-decoration` with the value `overline`. This leads to an overlining (not underlining!) of the relevant element, as you can see in see Figure 8.1.

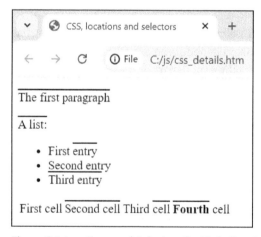

Figure 8.1 Locations and Selectors for Style Sheets

Here's the program:

```
<!DOCTYPE html><html lang="en-us">
<head>
    <meta charset="utf-8">
    <title>CSS, locations and selectors</title>
    <link rel="stylesheet" href="css_external.css">
    <style>
```

```
      p      {text-decoration:overline;}
      .over  {text-decoration:overline;}
      #idTop {text-decoration:overline;}
   <style>
</head>
<body>
   <p>The first paragraph</p>
   <p>A list:</p>
   <ul>
      <li>First <span class="über">entry</span></li>
      <li>Second entry</li>
      <li class="over">Third entry</li>
   </ul>

   <table>
      <tr>
         <td>First cell</td>
         <td id="idTop">Second cell</td>
         <td>Third <span style="text-decoration:overline;">cell</span></td>
         <td><b>Fourth</b> cell</td>
      </tr>
   </table>
</body></html>
```

Listing 8.1 css_details.htm File

You can use `<link rel="stylesheet" href="css_external.css">` to integrate external style sheets from the *css_external.css* file, which can be used by many files. We explain the content of the external file in the following text.

The style sheets from the `style` container, which is embedded in the file header, only apply to the current *css_details.htm* file. You could add explanatory comments to the style sheets, as in JavaScript between /* and */.

A single CSS property consists of a property and a value, separated by a colon and terminated with a semicolon. The entire CSS is enclosed in curly brackets.

Selectors assign the style sheets as follows:

- An HTML selector consists of the relevant HTML tag; here, it is p. As a result, the contents of all p containers are assigned an overline.

- A class selector is identified by a dot and a custom name for the class; here, it is .over. The rules for assigning names are similar to those for JavaScript variables (i.e., only letters, numbers, or an underscore—no number at the beginning). All HTML tags that use the over class have an overline.

285

- An ID selector starts with the hash character, #, followed by the ID; here, the complete selector is #idTop. This causes the content of the element with the idTop ID to have an overline.

Here's the content of the external file:

```
/* CSS specifications to implement */
b {text-decoration:overline;}
```

Listing 8.2 css_external.css File

The external file contains a comment and a style sheet with an HTML selector. This HTML selector has the effect that the contents of all b containers have an overline in all files that include this external file.

The two paragraphs are overlined due to the HTML selector in the file header. Within the list, there are two elements to which the over class is assigned: the span element in the first list entry and the entire third list entry. The assignment is made using the class attribute.

The second cell in the table has the idTop ID, and within the third cell, an inline specification is made using the style attribute. Curly brackets are not required here, and each CSS specification should end with a semicolon. In the fourth cell, there's a bold area that is overlined according to the external CSS.

8.1.2   Combinations

You can combine selectors and classes with each other, and you can see some examples of this in this section. In addition to the overline, you can use the CSS property font-weight with the value bold for bold formatting in the document. You can also use the CSS property font-style with the value italic for italics.

Here's the program:

```
<!DOCTYPE html><html lang="en-us">
<head>
    <meta charset="utf-8">
    <title>CSS, combinations</title>
    <style>
        p.over    {text-decoration:overline; font-weight:bold;}
        .slanted {font-style:italic;}
        li b      {text-decoration:overline;}
        div,td    {text-decoration:overline;}
    <style>
</head>
<body>
    <p class="over">The first paragraph</p>
```

```
   <p>The <i class="over">second</i> paragraph</p>
   <div>An additional line</div>
   <p>A <b>list</b>:</p>
   <ul>
      <li>First <b>entry</b></li>
      <li>Second entry</li>
   </ul>

   <table>
      <tr>
         <td>First cell</td>
         <td>Second cell</td>
      </tr>
   </table>
   <p class="over slanted">The last paragraph</p>
</body>
</html>
```

Listing 8.3 css_combinations.htm File

The p.over specification is a linked selector, and the two different selectors—in this, case the HTML selector and the class selector— must match. Only those paragraphs to which the over class is assigned are overlined and printed in bold. Several CSS properties are listed, separated by semicolons. Then follows a simple class selector: the slanted class.

The li b specification (without a dot in between) is a nested selector. The two selectors must also apply, but in nested form. Only those parts of the document are overlined that are inside a b container, which in turn is inside an li container.

You can line up several selectors one after the other, separated by commas. The following style sheets are used for all selectors, and both the contents of div containers and the contents of cells are overlined.

The first paragraph is overlined and printed in bold because it's also assigned the over class. In the second paragraph, nothing is overlined or bolded. The i container within this second paragraph is assigned the over class, but the entire paragraph is not. This is followed by an overlined div container.

No selector applies in the paragraph above the list. Although there's a b container, it's not within an li container. The situation is different for the second word of the first list entry: here, we have a b container within an li container.

There are two overlined td containers within the table.

Two classes are assigned to the last paragraph, separated by spaces. This means that both the style sheets of the over class and the style sheets of the slanted class are used.

You can see the result of the formatting in Figure 8.2.

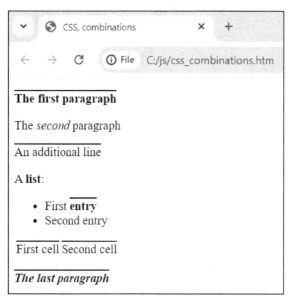

Figure 8.2 Combining Selectors

8.1.3   Cascading and Overlaying

Thanks to the many options of CSS, it's often the case that multiple style sheets apply to a particular element in the document. This behavior is desirable and increases flexibility. In addition to central formatting for an entire website, you can use local formatting without causing any contradictions.

The following rules apply when using multiple style sheets:

- If the CSS properties are different, they complement each other. This is also referred to as *cascading*.
- If the CSS properties are the same, the value of the definition closest to the element applies: an inline definition overlays an embedded definition, which in turn overlays an external definition.

Let's demonstrate this with an example:

```
<!DOCTYPE html><html lang="en-us">
<head>
    <meta charset="utf-8">
    <title>CSS, rules</title>
    <link rel="stylesheet" href="css_external.css">
    <style>
        b {text-decoration:underline;}
    <style>
```

```
</head>
<body>
    <p>The <b>first</b> paragraph</p>
    <p>The <b style="text-decoration:line-through;">second</b> paragraph</p>
    <p>The <b style="font-style:italic;">third</b> paragraph</p>
</body>
</html>
```

Listing 8.4 css_rules.htm File

First, the external style sheets from the *css_extern.css* file are integrated, as you already know from Section 8.1.1. This section states that bolded areas are overlined.

Embedded in the document header, the same property is defined for the same selector but overlaid with a different value. This means that within this document, areas in bold are underlined, as you can see in the first paragraph.

Within the second paragraph, a third definition is made for the same property, again for a bolded area. An overlay occurs again, and this area is crossed out.

In the third paragraph, there's a cascading of properties within the bold area. Another feature is added to the underlining: the italic font style.

The result is shown in Figure 8.3.

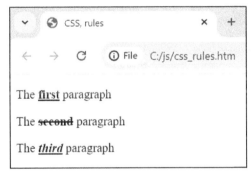

Figure 8.3 Cascading and Overlaying

8.2   Changing Properties

You define the CSS specifications for an HTML element using the style attribute. In JavaScript, you can change the CSS specifications using a property that is also named style. If you make a change repeatedly and in small steps using a time-controlled process, the result is an animation like in a movie.

In this section, we define the CSS properties for position, size, z-index, transparency, and color, and we then change and animate the associated subproperties of the style

289

property. The *z-index* is the position of an element in the *z* direction (i.e., toward or away from the viewer).

8.2.1 Position

The CSS position, top, and left properties are used to define the position of an element as follows:

- The CSS position property stands for the type of positioning. The absolute value is often used in this context, and it stands for an absolute positioning that refers to the edge of the document.

- The CSS top and left properties define the distance between the top left corner of the positioned element and the top left corner of the document. The value is usually given in the px unit, for pixels.

Let's now take a look at the document in Figure 8.4.

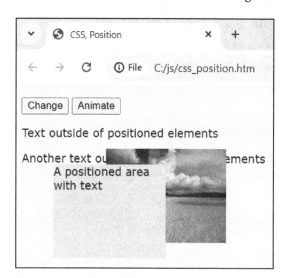

Figure 8.4 Positioned Elements

There are two positioned elements in the document: an image and a div container with text against a dark background.

Here's the lower part of the program with the structure of the document:

```
...
<body>
    <p><input id="idChange" type="button" value="Change">
        <input id="idAnimate" type="button" value="Animate"></p>
    <p>Text outside of positioned elements</p>
    <p>Another text outside of positioned elements</p>
```

```
<img src="im_paradise.jpg" alt="Paradise" id="idImage"
    style="position:absolute; top:80px; left:120px;">
<div style="position:absolute; top:100px; left:50px;
        width:150px; height:120px; background-color:#e0e0e0;">
  A positioned area with text</div>

<script>
    const image = document.getElementById("idImage");
    document.getElementById("idChange").addEventListener("click", change);
    document.getElementById("idAnimate").addEventListener("click",
        function() { reference = setInterval(animate, 20); } );
</script>
</body></html>
```

Listing 8.5 css_position.htm File, Lower Part

The image has the style attribute. Without the CSS position property, the CSS top and left properties would have no effect. The div container also has values for the CSS width and height properties for the width and height, respectively, and background-color for the background color.

Positioned elements are independent of nonpositioned elements in the document. They can conceal other elements. The div container has been defined in the code after the image, and for this reason and also because of its position and size, the div container partially conceals the image.

The image variable references the image. The first event handler connects clicking on the **Change** button with the change() function, and clicking on the **Animate** button starts a time-controlled process that moves the image in an animated manner.

Now, here's the upper part of the program with the JavaScript code:

```
... <head> ...
  <script>
    function change()
    {
        image.style.top = "130px";
        image.style.left = "220px";
    }

    let reference, imTop = 80, imLeft = 120;

    function animate()
    {
        if (imTop >= 130)
            clearInterval(reference);
        else
```

291

```
        {
            imTop++;
            imLeft += 2;
            image.style.top = imTop + "px";
            image.style.left = imLeft + "px";
        }
    }
  </script>
</head>
...
```

Listing 8.6 css_position.htm File, Upper Part

The change() function gives new values to the top and left subproperties of the style property. These values are located within a string and contain the px unit. In this way, the image gets moved from the default position to another position.

The reference variable references the time-controlled process. The imTop and imLeft variables contain the two numerical values for the imTop and imLeft subproperties, and their initial values correspond to the values agreed in the style attribute of the image.

The current position is checked in the animate() function. The value 130 for the imTop subproperty corresponds to the final position, and once this is reached, the time-controlled process terminates. If this hasn't yet been reached, the image will get moved down by 1 pixel and to the right by 2 pixels each time the animate() function is run. The px unit is appended to the numerical value.

After the position is changed, the image is at the bottom right, as you can see in Figure 8.5.

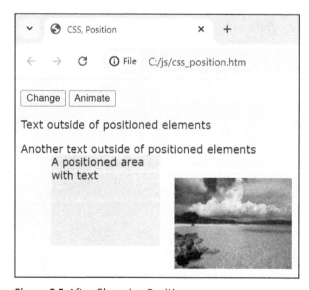

Figure 8.5 After Changing Position

8.2.2   Size

The CSS `width` and `height` properties are used to define the size of an element. There's a positioned image in the document (see Figure 8.6).

After the size is changed, the image looks as shown in Figure 8.7.

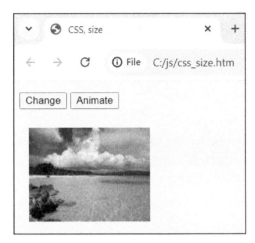

Figure 8.6 Positioned Element

Figure 8.7 After Changing Size

Here's the first part of the program:

```
...
<body>
    <p><input id="idChange" type="button" value="Change">
        <input id="idAnimate" type="button" value="Animate"></p>
```

```
    <img src="im_paradise.jpg" alt="Paradise" id="idImage"
        style="position:absolute; top:60px; left:20px;
             width:160px; height:120px;">
  <script>
    const image = document.getElementById("idImage");
    document.getElementById("idChange").addEventListener("click", change);
    document.getElementById("idAnimate").addEventListener("click",
        function() { reference = setInterval(animate, 20); } );
  </script>
</body></html>
```

Listing 8.7 css_size.htm File, Lower Part

The image is displayed in the size of 160 × 120 pixels using the CSS properties width and height. In this case, this also corresponds to the original size. The reference to the image and the event handlers corresponds to that in Section 8.2.1.

Now, here's the upper part of the program with the JavaScript code:

```
... <head> ...
  <script>
    function change()
    {
        image.style.width = "240px";
        image.style.height = "180px";
    }

    let reference, imWidth = 160, imHeight = 120;

    function animate()
    {
        if (imWidth >= 240)
            clearInterval(reference);
        else
        {
            imWidth++;
            imHeight += 0.75;
            image.style.width = imWidth + "px";
            image.style.height = imHeight + "px";
        }
    }
  </script>
</head>
...
```

Listing 8.8 css_size.htm File, Upper Part

The change() function gives new values to the imWidth and imHeight subproperties of the style property.

For the animation, the imWidth and imHeight variables contain the numerical values for the two subproperties. In the animate() function, the imWidth variable is changed in increments of 1 up to the value 240. To maintain the aspect ratio of the image, the imHeight variable is only changed in steps of 0.75.

8.2.3 Position in the Z Direction

You can influence the position of positioned elements in the z direction by using the z-index CSS property. If you make no entry, the value of z-index is 0, but in this case, the element is at least in front of an element that doesn't get positioned.

An element with a higher value is in front of an element with a lower value in the z direction, and an element with a negative value is even behind an element that is not positioned.

You can see an example in Figure 8.8.

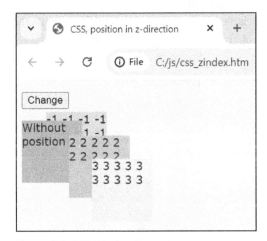

Figure 8.8 CSS Z-Index Property

The document contains three div containers positioned in different colors that partially cover each other. The z-index CSS property has different values: -1, 2, or 3, and this results in the sequence in the z direction. The div container with the value -1 for the z-index CSS property is partially covered by a nonpositioned element.

Here's the first part of the program:

```
...
<body>
    <p><input id="idChange" type="button" value="Change"></p>
    <p style="background-color:#c0c0c0; width:80px; height:80px;">
```

```
    Without position</p>
  <div style="position:absolute; top:100px; left:100px; width:80px;
      height:80px; z-index:3; background-color:#f0f0f0;">
    3 3 3 3 3 3 3 3 3 3</div>
  <div id="idArea" style="position:absolute; top:70px; left:70px;
      width:80px; height:80px; z-index:2; background-color:#d0d0d0;">
    2 2 2 2 2 2 2 2 2 2</div>
  <div style="position:absolute; top:40px; left:40px; width:80px;
      height:80px; z-index:-1; background-color:#e0e0e0;">
    -1 -1 -1 -1 -1 -1 -1 -1 -1 -1 </div>
  <script>
    const area = document.getElementById("idArea");
    document.getElementById("idChange").addEventListener("click", change);
  </script>
</body></html>
```

Listing 8.9 css_zindex.htm File, Lower Part

The order in which the div containers are defined is irrelevant, as they all have a value for the z-index CSS property. The area variable references the div container with the value 2 for the z-index CSS property, and a click on the **Change** button calls the change() function.

Now, here's the upper part of the program with the JavaScript code:

```
... <head> ...
  <script>
    function change()
    {
      area.style.zIndex = 4;
      area.firstChild.nodeValue = "4 4 4 4 4 4 4 4 4 4";
    }
  </script>
</head>
...
```

Listing 8.10 css_zindex.htm File, Upper Part

The change() function gives the new value 4 to the zIndex subproperty of the style property. This means that this div container is at the very front and may conceal other div containers and nonpositioned elements of the program (see Figure 8.9). The text content of the div container is changed accordingly.

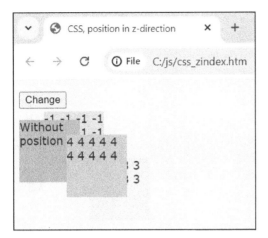

Figure 8.9 After Changing Position in Z Direction

8.2.4   Transparency

You can set the transparency of an element by using the opacity CSS property. You can enter any value between 0.0 and 1.0. The default value of 1.0 means that the element is opaque (i.e., not transparent), which means that an element behind it in the z direction will not be recognizable. The closer the value of opacity gets to 0.0, the more transparent it becomes: at 0.0, it's completely transparent and therefore invisible.

In Figure 8.10, you can see writing in a div container for which no transparency has been set. In addition, three other div containers appear for which different values have been set for transparency.

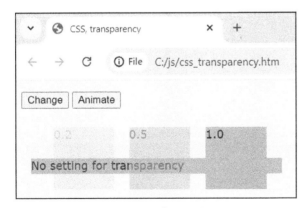

Figure 8.10 Transparency with Opacity

Here's the first part of the program:

```
...
<body>
    <p><input id="idChange" type="button" value="Change">
```

```
      <input id="idAnimate" type="button" value="Animate"></p>
   <div id="idArea" style="top:100px; left:20px; width:330px;
      height:20px;">No setting for transparency</div>
   <div style="top:60px; left:50px;  opacity:0.2;">0.2</div>
   <div style="top:60px; left:150px; opacity:0.5;">0.5</div>
   <div style="top:60px; left:250px; opacity:1.0;">1.0</div>
   <script>
      const area = document.getElementById("idArea");
      document.getElementById("idChange").addEventListener("click", change);
      document.getElementById("idAnimate").addEventListener("click",
         function() { reference = setInterval(animate, 20); } );
   </script>
</body></html>
```

Listing 8.11 css_transparency.htm File, Lower Part

The three square div containers are given the values 0.2, 0.5, and 1.0 for the opacity CSS property.

The area variable references the div container with the text No setting for transparency, and clicking the **Change** and **Animate** buttons calls the change() and animate() functions, respectively.

Now, here's the upper part of the program with the JavaScript code:

```
... <head> ...
   <style>
      div {position:absolute; width:80px; height:80px;
         background-color:#c0c0c0;}
   <style>
   <script>
      function change()
      {
         area.style.opacity = 0.4;
      }

      let reference, value = 1.0;

      function animate()
      {
         if (value <= 0.0)
           clearInterval(reference);
         else
         {
           value -= 0.01;
           area.style.opacity = value;
```

```
            area.firstChild.nodeValue = "Transparency " + value.toFixed(2);
        }
    }
  </script>
</head>
...
```

Listing 8.12 css_transparency.htm File, Upper Part

First, some common CSS properties are set for the div containers. The change() function assigns the value 0.4 to the opacity subproperty of the style property (see Figure 8.11).

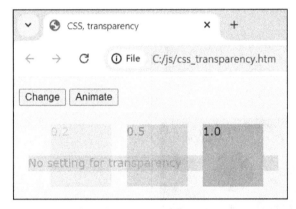

Figure 8.11 After Change in Transparency

For the animation, the value variable contains the numerical value for this subproperty. The animate() function changes the value variable from 1.0 to 0.0 in increments of 0.01. At the beginning of the animation, the element is completely opaque, and at the end, it's entirely transparent. The current transparency is displayed as the text content of the div container.

8.2.5   Visibility

A transparency of 0.0 makes an element invisible, and you can also influence the visibility of the element by using the visibility CSS property. The visible value makes an element visible, while the hidden value makes it invisible. An invisible element occupies space in the document but doesn't respond to events such as mouseover or mouseout.

The document in the following example (see Figure 8.12) contains two images. The first image is invisible at the beginning. When the mouse is positioned over a visible image, the other image is made invisible. If the mouse leaves the area of a visible image, the other image becomes visible (see Figure 8.13).

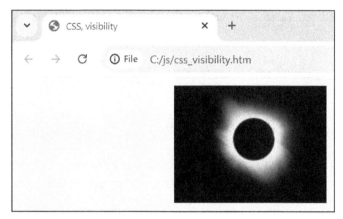

Figure 8.12 Visibility Using Visibility Property

Figure 8.13 After Leaving Area of Second Image

Here's the program:

```
...
<body>
   <img src="im_tree.jpg" alt="Tree" id="idTree" style="visibility:hidden;">
   <img src="im_solar_eclipse.jpg" alt="SolarEclipse" id="idSolarEclipse">
   <script>
      const tree = document.getElementById("idTree");
      const se = document.getElementById("idSolarEclipse");
      tree.addEventListener("mouseover",
         function() { se.style.visibility = "hidden"; });
      tree.addEventListener("mouseout",
         function() { se.style.visibility = "visible"; });
      se.addEventListener("mouseover",
         function() { tree.style.visibility = "hidden"; });
      se.addEventListener("mouseout",
```

```
        function() { tree.style.visibility = "visible"; });
    </script>
</body></html>
```

Listing 8.13 css_visibility.htm File

The program creates links to the two images. As a result of the mouseover and mouseout events, the value of the visibility subproperty of the style property of the other image is set using an anonymous function.

8.2.6 Color

You can set the background color of an element by using the background-color CSS property, and you can set the text color of an element by using the color CSS property. There are several ways to choose a color, but especially for an animation, we recommend that you assign the value using the rgb() formula. You can use values from 0 to 255 for the red, green, and blue components of the color.

In Figure 8.14, you can see light-gray writing in a dark-gray div container.

Figure 8.14 Text Color and Background Color

Here's the first part of the program:

```
...
<body>
    <p><input id="idChange" type="button" value="Change">
        <input id="idAnimate" type="button" value="Animate"></p>
    <div id="idArea" style="position:absolute; top:50px; left:50px;
            width:100px; height:100px; color:rgb(192,192,192);
            background-color:rgb(64,64,64);">Light gray on dark gray</div>
    <script>
        const area = document.getElementById("idArea");
        document.getElementById("idChange").addEventListener("click", change);
```

```
    document.getElementById("idAnimate").addEventListener("click",
        function() { reference = setInterval(animate, 20); } );
  </script>
</body></html>
```

Listing 8.14 css_color.htm File, Lower Part

The values 192/192/192 correspond to light gray, and the values 64/64/64 correspond to dark gray. The div container contains information on the current color setting. The area variable references the positioned div container, clicking the **Change** button calls the change() function, and clicking the **Animate** button calls an anonymous function that calls the method setInterval().

Now, here's the upper part of the program with the JavaScript code:

```
... <head> ...
  <script>
    function change()
    {
        area.style.color = "rgb(64,64,64)";
        area.style.backgroundColor = "rgb(192,192,192)";
        area.firstChild.nodeValue = "Dark gray on light gray";
    }

    let reference, red = 64, green = 64, blue = 64;
    function animate()
    {
        if (red >= 192)
        {
          clearInterval(reference);
          area.firstChild.nodeValue = "Dark gray on light gray";
        }
        else
        {
          red++;
          green++;
          blue++;
          area.style.color = "rgb(" + (256-red)
              + "," + (256-green) + "," + (256-blue) + ")";
          area.style.backgroundColor =
              "rgb(" + red +", "+ green +", "+ blue + ")";
        }
    }
```

```
  </script>
</head>
...
```

Listing 8.15 css_color.htm File, Upper Part

The change() function gives new values to the color and backgroundColor subproperties of the style property (see Figure 8.15). It assigns these values using the rgb() formula within a string. The div container also has new content: the new color setting.

For the animation, the red, green, and blue variables contain the three numerical values for the color components. The animate() function changes these proportions in increments of 1 from the value 64 to the value 192. These values are used directly to change the background color, and the text color changes in reverse order. Since three variables are used, you could set any start and end colors for the animation.

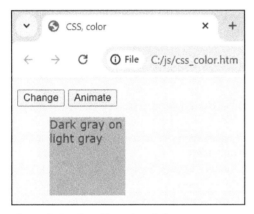

Figure 8.15 After Changing Colors

8.3   Additional Options

This section contains additional programs that use the properties described previously.

> **Note**
> The names of the subproperties of the style property are very similar to the names of the associated CSS properties. If the name of the CSS property contains a hyphen, then it gets omitted and the letter that follows is capitalized. Here are two examples: border-width (total frame width) becomes borderWidth, and border-top-width becomes borderTopWidth. If the value of a CSS property consists of more than one numerical value, then it's noted as a character string in JavaScript.

8.3.1   Transparency when Changing Images

In the following program, two images are positioned in the same place. The first image is displayed without setting the transparency (see Figure 8.16), and the second image has a transparency of 0.0 (i.e., it's initially completely transparent).

A click on the **Animate** button changes the values for transparency. The first image becomes increasingly transparent, and the second image becomes increasingly opaque (see Figure 8.17). In this way, an animated image change is carried out—and in the end, only the second image is visible (see Figure 8.18).

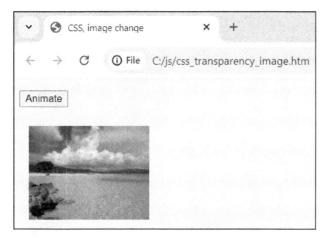

Figure 8.16 First Image Visible, Second Image Transparent

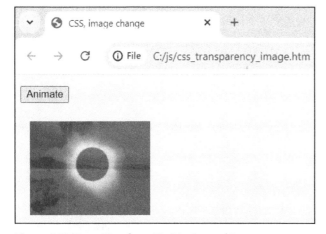

Figure 8.17 Transition from First to Second Image

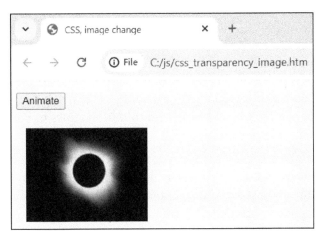

Figure 8.18 Second Image Visible, First Image Transparent

Here's the program:

```
...  <head> ...
   <script>
      let reference, value = 1;
      function animate()
      {
         if (value <= 0)
            clearInterval(reference);
         else
         {
            value -= 0.01;
            pa.style.opacity = value;
            se.style.opacity = 1 - value;
         }
      }
   </script>
</head>
<body>
   <p><input id="idAnimate" type="button" value="Animate"></p>
   <img src="im_paradise.jpg" alt="Paradise" id="idParadise"
      style="position:absolute; top:60px; left:20px;">
   <img src="im_solar_eclipse.jpg" alt="SolarEclipse" id="idSolarEclipse"
      style="position:absolute; top:60px; left:20px; opacity:0.0;">
   <script>
      const pa = document.getElementById("idParadise");
      const se = document.getElementById("idSolarEclipse");
      document.getElementById("idAnimate").addEventListener("click",
         function() { reference = setInterval(animate, 20); } );
```

```
    </script>
</body></html>
```

Listing 8.16 css_transparency_image.htm File

Both images are positioned at 60 × 20 pixels and have a size of 160 × 120 pixels. In the second image, the value of the `opacity` CSS property is set to 0.0, and links to both images and an event handler for the button are set up.

The `value` variable is used to set the `opacity` subproperty of the `style` property of both images. The value of the variable starts at 1.0, decreases in increments of 0.01, and ends at 0.0. The value is assigned directly to the first image, and in the second image, the transparency changes in the opposite direction. The difference 1 - `value` is therefore calculated beforehand.

8.3.2   Visibility of a Menu

We use the visibility to dynamically display a menu with a submenu. In Figure 8.19, you can see the menu in its initial state. If the mouse is positioned over one of the menu items, the corresponding submenu becomes visible (see Figure 8.20), and it's only hidden again when the mouse leaves the area of the associated menu or the displayed submenu.

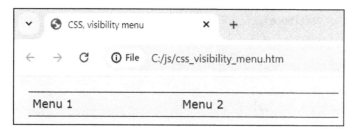

Figure 8.19 Menu in Start State

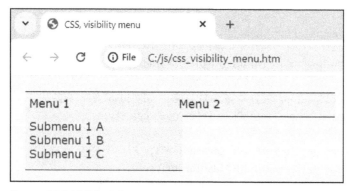

Figure 8.20 Visible Submenu

Here's the sample program:

```
... <head> ...
   <script>
      function subOn(x)
      {
         if(x==1)
            sub1.style.visibility = "visible";
         else
            sub2.style.visibility = "visible";
      }

      function subOff(x)
      {
         if(x==1)
            sub1.style.visibility = "hidden";
         else
            sub2.style.visibility = "hidden";
      }
   </script>
   <style>
      div {position:absolute; width:200px; height:20px; padding:5px;
         border-top:solid 1px #000000; border-bottom:solid 1px #000000;
         background-color:#f0f0f0;}
      .sub    {visibility:hidden; height:60px; border-top-width:0px;}
      a:link  {color:#000000; text-decoration:none;}
      a:hover {background-color:#d0d0d0;}
   <style>
</head>
<body>
   <div id="idMenu1" style="top:20px; left:20px;">Menu 1</div>
   <div id="idMenu2" style="top:20px; left:220px;">Menu 2</div>
   <div id="idSub1" class="sub" style="top:50px; left:20px;">
      <a href="#">Submenu 1 A</a><br>
      <a href="#">Submenu 1 B</a><br>
      <a href="#">Submenu 1 C</a>
   </div>
   <div id="idSub2" class="sub" style="top:50px; left:220px;">
      <a href="#">Submenu 2 A</a><br>
      <a href="#">Submenu 2 B</a><br>
      <a href="#">Submenu 2 C</a>
   </div>
   <script>
      const menu1 = document.getElementById("idMenu1");
      menu1.addEventListener("mouseover", function() { subOn(1); });
```

8

```
        menu1.addEventListener("mouseout",  function() { subOff(1); });
        const menu2 = document.getElementById("idMenu2");
        menu2.addEventListener("mouseover", function() { subOn(2); });
        menu2.addEventListener("mouseout",  function() { subOff(2); });
        const sub1 = document.getElementById("idSub1");
        sub1.addEventListener("mouseover", function() { subOn(1); });
        sub1.addEventListener("mouseout",  function() { subOff(1); });
        const sub2 = document.getElementById("idSub2");
        sub2.addEventListener("mouseover", function() { subOn(2); });
        sub2.addEventListener("mouseout",  function() { subOff(2); });
    </script>
</body></html>
```

Listing 8.17 css_visibility_menu.htm File

Both the main menu items and the submenus are in div containers with the following properties:

- A size of 200 × 20 pixels
- One upper and one lower edge
- A light-gray background color
- An inner distance of five pixels to the edge

The properties of the submenus are added or overwritten using the sub class: the height is changed to 60 pixels, the top border is removed, and most importantly, they are invisible.

The menu items in the submenus are hyperlinks that can be used to call other pages, for example. For these hyperlinks, the following definitions are made: text color black, no underlining. If the mouse is positioned over one of the hyperlinks, the background color changes.

Menus and submenus are positioned to match each other in the document itself, and the sub class is assigned to the submenus.

Hovering over a main menu item with the mouse calls the subOn() function, and the value 1 or the value 2 is transmitted as a parameter. Within the subOn() function, the visibility subproperty of the style property of the corresponding submenu is set to visible.

Conversely, leaving a main menu item with the mouse leads to the function subOff() being called with the same parameters. Within the subOff() function, the visibility subproperty of the style property of the corresponding submenu is set to hidden.

Hovering or leaving one of the submenus leads to the same results. Otherwise, the relevant submenu disappears as soon as the mouse is moved from the associated main menu item to the submenu.

The positions are selected in such a way that the submenus slightly overlap the main menu items. Otherwise, the submenus would become invisible as soon as the mouse left the associated main menu item.

8.3.3   Animated Throw

In the downloadable materials for this book, you can find the *css_throw.htm* program with many explanatory comments as a bonus. This program lets you simulate throwing a ball at a target.

For this purpose, you can adjust the angle of the throw to the ground and the ball's speed at the start of the throw. The program takes into account the influence of gravity so that the ball's trajectory resembles a parabola. You can also place the target at a fixed or a random location (see Figure 8.21).

The trajectory of the ball in the *x* and *y* directions after a time is determined by two physical formulas (see also *https://en.wikipedia.org/wiki/Projectile_motion*):

$$sx(t) = s_0 + v_0 * \cos(w) * t$$
$$sy(t) = s_0 + v_0 * \sin(w) * t + 0.5 * a_0 * t * t$$

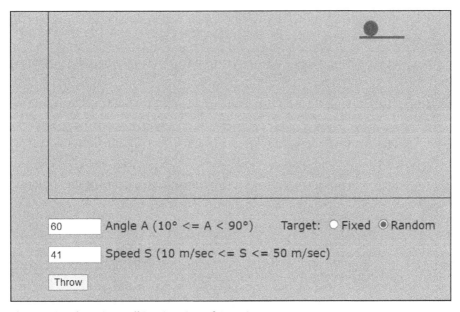

Figure 8.21 Throwing Ball in Direction of Target

Here's more information about the individual terms in the physical formulas:

- s_0 is the *trajectory* at the start of the throw. This trajectory corresponds to the coordinates of the starting point.

- v_0 is the *speed* at the start of the throw (i.e., the speed set in the input field). The components of the speed in the x direction and in the y direction can be calculated depending on the angle w using the cosine and the sine. The angle is set in the input field.

- a_0 is the *acceleration* at the start of the throw, which is dependent on the gravity. It only exists in the y direction.

The expressions of the formulas are given factors in the program so that the movement looks more or less natural.

Chapter 9

Two-Dimensional Graphics and Animations Using SVG

You can use SVG to display two-dimensional vector graphics and animations in the browser.

Scalable Vector Graphics (SVG) is an XML-based language format that you can use to create two-dimensional vector graphics. You can create a *vector graphic* using a description; it requires little storage space, and you can easily change it without any loss of quality (e.g., with regard to its position, size, or rotation).

SVG can contain animations and be displayed in modern browsers without additional programs. Using JavaScript and the DOM you know from Chapter 5, you can access, change, or add to individual elements of SVG.

You can embed SVG code directly into your HTML code or save it in a separate SVG file, which provides a (possibly animated) image that you can insert as a multimedia object anywhere in different HTML documents using the object HTML tag. This chapter shows only some of the extensive possibilities of SVG.

The programs in this chapter are based on the SVG 1.1 standard, which the World Wide Web Consortium (W3C) published in August 2011. A draft for the SVG 2 standard has been available since September 2016, although it has only been partially implemented by the individual browsers to date.

9.1 Creating an SVG File

In this section, we create a first SVG file and embed it in an HTML document. The SVG has a frame for clarification and contains a black rectangle at a specific position (see Figure 9.1).

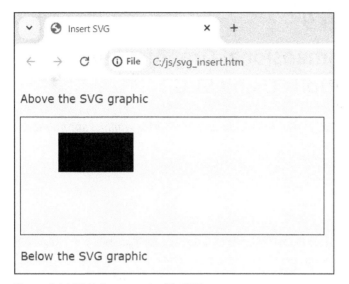

Figure 9.1 HTML Document with SVG

Here's the code of the SVG file:

```
<svg xmlns="https://www.w3.org/2000/svg" width="400" height="150">
    <!-- Rectangle -->
    <rect x="50" y="20" width="100" height="50" />
</svg>
```

Listing 9.1 svg_insert.svg File

The svg tag introduces the SVG code. According to the DOM from Chapter 5, this tag is the root node in which all child nodes (the graphic elements) are embedded. The child nodes can in turn have child nodes and so on, and each node can also have attribute nodes.

The xmlns attribute uses the specified value to define the XML *namespace* from which the SVG elements used originate, and the width and height attributes determine the width and height of the graphic in pixels. Comments are created as in HTML, using <!-- and -->.

If an SVG element has no subelement (i.e., no child node), no end tag is required. In this case, the start tag must be terminated with />. A subelement could, for example, be an animation that references the SVG element.

The rect tag creates a rectangle, which is positioned using the *x* and *y* coordinates of its top left-hand corner. The values refer to the top left-hand corner of the graphic, and the size of the rectangle is defined using the width and height attributes. If no further information is provided, an SVG element is filled and appears black.

Now, here's the code of the HTML file:

```
...
<body>
   <p>Above the SVG graphic</p>
   <p><object data="svg_insert.svg" type="image/svg+xml"
        width="400" height="150" style="border:1px solid black;">
     Here is an SVG graphic
   </object></p>
   <p>Below the SVG graphic</p>
</body></html>
```

Listing 9.2 svg_insert.htm File

You can insert the SVG file into an HTML document as a multimedia object by using the `object` HTML tag. You could also use the `img` HTML tag, but then, existing animations would not be executed.

The `data` attribute references the SVG file, and the `type` attribute refers to the type of file. The `width` and `height` attributes define the size of the multimedia object within the HTML document, which here has a thin black frame for clarity. The "Here is an SVG graphic" text is only visible if the browser can't display the SVG graphic.

9.2 Basic Shapes

The first basic shapes are displayed in the *svg_basic_shapes.htm* file: rectangles (see Figure 9.2), circles, and ellipses (see Figure 9.3) as well as lines, polylines, and polygons (see Figure 9.4).

The SVG code is directly embedded in the HTML code in this and the other examples in this chapter. You can save the SVG code of all graphics with the knowledge you gained in Section 9.1 but also in separate SVG files.

9.2.1 Rectangles

Here's the code for the rectangles from Figure 9.2:

```
...
<body>
   <svg xmlns="https://www.w3.org/2000/svg" width="600"
        height="300" style="border:1px solid black;">
     <rect x="50"  y="20" width="100" height="50" />
     <rect x="200" y="20" width="100" height="50" rx="10" ry="10"
        fill="#bbbbbb" stroke="#000000" stroke-width="2" />
     <rect x="350" y="20" width="100" height="50" />
```

```
    <rect x="355" y="25" width="50"  height="25" fill="#ffffff" />
...
  </svg>
</body></html>
```

Listing 9.3 svg_basic_shapes.htm File, Structure, and Rectangles

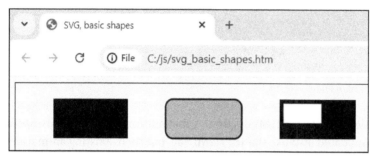

Figure 9.2 Rectangles

The SVG container is embedded in the body of the HTML document and can be surrounded by other HTML tags and text. A graphic size of 600 × 300 pixels is selected, and the frame line is assigned directly to the SVG element.

We described how to create a simple black rectangle in Section 9.1.

You give the second rectangle rounded corners by using the rx and ry attributes. You can make the rounding different in the *x* and *y* directions, and you can use the fill attribute to define a fill color for SVG elements. The color in our example is gray, but the default color is black.

The stroke attribute defines the color of the frame line (which is black here) for SVG elements and also ensures that it gets displayed. The frame line of an SVG element has a default stroke width of one pixel; you can select a value of two pixels by using the stroke-width attribute.

An SVG element can completely or partially cover another SVG element that was created previously in the code. The small white rectangle covers part of the large black rectangle.

9.2.2 Circles and Ellipses

Now, here's the code for the circles and the ellipse from Figure 9.3:

```
...
    <circle cx="80"  cy="130" r="30" />
    <circle cx="180" cy="130" r="30" stroke="#000000"
      stroke-width="2" fill="#bbbbbb" />
    <circle cx="280" cy="130" r="30" />
```

```
<circle cx="295" cy="130" r="30" fill="#ffffff" />

<ellipse cx="400" cy="130" rx="60" ry="30"/>
...
```

Listing 9.4 svg_basic_shapes.htm File, Circles, and Ellipses

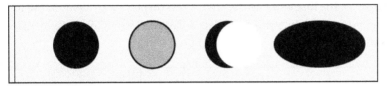

Figure 9.3 Circles and Ellipses

The circle tag creates a circle, and you position it using the cx and cy attributes for the x and y coordinates of its center. You define the size of the circle by using the r attribute for the radius, and if you make no further entries, this SVG element will also be filled and black, as you can see in the first circle.

The fill color and the color and line width of the frame line are defined for the second circle.

You create the "crescent moon" by covering the right part of a black circle with a white circle.

The ellipse tag creates an ellipse. As with the circle, you carry out positioning by using the cx and cy attributes for the coordinates of the center of the ellipse. An ellipse has two radii, and you determine the radius in the x direction by using the rx attribute and the radius in the y direction by using the ry attribute. If both radii are the same, then the ellipse is actually a circle.

9.2.3 Lines, Polylines, and Polygons

Here's the code of the line as well as the polylines and the polygons from Figure 9.4:

```
...
<line x1="20" y1="190" x2="20" y2="250"
    stroke="#000000" stroke-width="2" />

<polyline points=" 50,250  80,190 110,250" />
<polyline points="150,250 180,190 210,250"
   stroke="#000000" stroke-width="2" fill="none" />
<polyline points="250,250 280,190 310,250"
   stroke="#000000" stroke-width="2" fill="#bbbbbb" />
```

```
<polygon points="350,250 380,190 410,250"
    stroke="#000000" stroke-width="2" fill="#bbbbbb" />
```
...

Listing 9.5 svg_basic_shapes.htm File, Lines, Polylines, and Polygons

The line tag creates a line that starts at x1 and y1 and ends at x2 and y2. The stroke attribute defines the color of the line and ensures that it gets displayed. By default, the line has a stroke width of one pixel, but you can select a value of two pixels here by using the stroke-width attribute.

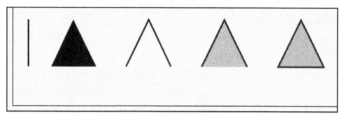

Figure 9.4 Lines, Polylines, and Polygons

A *polyline* is a set of lines, and you connect individual points (each of which, you specify with a pair of *x* and *y* coordinates) to each other to create the polyline. A *polygon* is also a set of lines, with the difference being that the last point is connected to the first point so that the polyline is closed. In this program, you always connect three points to each other to form a triangle that is either a polyline or a polygon.

The polyline tag creates a polyline, and the polygon tag creates a polygon. You position the individual points by using the points attribute, and you specify the pairs of *x* and *y* coordinates one after the other. Spaces and commas are permitted as separators, and I recommend separating the *x* and *y* coordinates of a pair with a comma for clarity. You should separate the different pairs with a space.

If no further information is entered, this SVG element is also filled and black, as you can see in the first polyline.

If the polyline or polygon is not supposed to be filled, you must specify the fill attribute with the none value. In this case, you should use the stroke attribute so that the polyline or polygon is visible in the same way as the second polyline.

The third polyline has a gray filling and, independently of this, a black line.

The polygon on the far right is constructed like the neighboring third polyline, but the third point is automatically connected to the first point so that the line is closed.

If a polyline only includes two points or if all points of a polyline lie on a line one behind the other, no area is created. In this case, there can be no filling, and if you don't enter a value for the stroke attribute for such a line, it won't be visible at all. This also applies to paths (Section 9.3).

9.3 Paths

A *path* is a line that can consist not only of straight lines but also of curves. A path can also be interrupted, and you create the various line shapes using a *drawing pen* that moves across the graphic. You can create all basic shapes from paths.

You can create the elements of a path using absolute or relative coordinates:

- Absolute coordinates are introduced by a specification in uppercase letters, and as with the previously known elements, they refer to the zero point of the graphic at the top left.

- Relative coordinates are introduced by a specification in lowercase letters, and they refer to the previous element of the path. They are often easier to handle than absolute coordinates.

You can fill paths with a color, just like the basic shapes. The *svg_path.htm* file contains some examples.

9.3.1 Filled Paths

Let's first take a look at the code of the four filled paths from Figure 9.5:

```
...
<body>
    <svg xmlns="https://www.w3.org/2000/svg" width="600px"
        height="250px" style="border:1px solid black;">
      <path d="M  50,20 L 80,50 L 80,80 L 50,80" />
      <path d="M 120,20 L 150,50 V 80 H 120" />
      <path d="M 190,20 l 30,30 l 0,30 l -30,0" />
      <path d="M 260,20 l 30,30 v 30 h -30" />
...
    </svg>
</body></html>
```

Listing 9.6 svg_path.htm File, Filled Paths

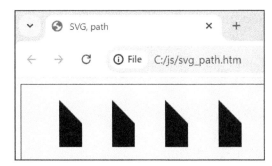

Figure 9.5 Filled Paths

The path tag creates a path, and you position the individual points using the d attribute. The following specifications can appear in the value for the d attribute:

- M, for *move to*: This specification moves the drawing pen to the next pair of absolute coordinates without drawing a line. It can be used to achieve an interruption in the line.

- m: This specification is like M, but it moves the drawing pen to the next pair of relative coordinates.

- L, for *line to*: This specification draws a line to the next pair of absolute coordinates.

- l: This specification is like L, but it draws a line to the next pair of relative coordinates.

- V, for *vertical*: This specification draws a vertical line to the next absolute *y* coordinate.

- v: This specification is like V, but it draws a vertical line to the next relative *y* coordinate. A positive value moves the drawing pen downward, and a negative value moves it upward.

- H, for *horizontal*: This specification draws a horizontal line to the next absolute *x* coordinate.

- h: This specification is like H, but it draws a horizontal line to the next relative *x* coordinate. A positive value moves the drawing pen to the right, and a negative value moves it to the left.

The four filled paths have the same appearance but are structured differently:

- The first path uses absolute coordinates using M or L.
- The second path uses absolute coordinates using M, L, V, and H.
- The third path uses M for the starting point and l for the subsequent relative coordinates.
- The fourth path uses M for the starting point and l, v, and h, for the subsequent relative coordinates.

The fourth path is easier to read than the previous paths. If no further information is entered, this SVG element is also filled and black.

9.3.2 Groups and Paths

Here's the code of the group and the three unfilled paths from Figure 9.6:

```
...
    <g stroke="#000000" stroke-width="2" fill="none">
        <path d="M 330,20 l 30,30 v 30 h -30" />
        <path d="M 400,20 l 30,30 v 30 h -30 z" />
        <path d="M 470,20 l 30,30 m 0,30 h -30" />
```

```
. . .
    </g>
. . .
```

Listing 9.7 svg_path.htm File, Unfilled Paths, and Group

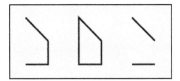

Figure 9.6 Unfilled Paths

You use the g tag to group elements to which you want to assign common attributes and values. The none value for the fill attribute indicates that the three unfilled paths and other elements in this group are not filled. You use the stroke attribute to make the polyline visible. The three paths have a similar structure, but there are small differences:

- The first path uses M for the starting point and l, v, and h for the subsequent relative coordinates.

- The second path ends with the z specification, and you use this to close the polyline (i.e., to make a connection from the last element to the starting point).

- In the third path, you use the m specification once instead of the v specification. This ensures that the pen moves to the next point without drawing a line and that the path is therefore shown with an interruption. However, it's still a logically connected path.

9.3.3 Paths with Curves

Paths with curves have a complex structure. There are elliptical arcs as well as square and cubic Bezier curves. In this introduction, I will only deal with elliptical arcs.

The following is the code of the path from Figure 9.7, which contains a total of three curves:

```
. . .
    <g stroke="#000000" stroke-width="2" fill="none">
. . .
        <path d="M 50,120
                v 60
                a 30,30 0 0 0 60,0
                v -30
                a 30,30 0 0 1 30,-30
                h 60
```

```
                        a 30,30 0 1 1 -30,30" />
        </g>
...
```

Listing 9.8 svg_path.htm File, Path with Curves

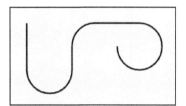

Figure 9.7 Path with Curves

The individual entries for the d attribute are placed one below the other for clarity. The A or a specification creates an elliptical arc. A total of seven values are required:

- The two radii of the ellipse in the *x* and *y* directions (see also Section 9.2.2). To simplify matters, the values for the two radii of all three ellipses are chosen to be the same, so that the curves are circular.

- The rotation of the arc around the *x* direction, given in degrees. This specifies by how many degrees the arc is tilted around its *x* axis. To simplify matters, the value 0 is selected for all three curves so that none of the curves are tilted.

- A value that determines whether the arc is drawn over a short path (value = 0) or a long path (value = 1) from the starting point to the end point. Here, the long path is only selected for the third arc so that it's drawn across three quarters of the arc.

- A value that defines the direction of rotation (i.e., whether the arc runs anticlockwise [value = 0] or clockwise [value = 1] from the starting point to the end point. Here, the second and third curves run clockwise.

- The absolute (arc with A) or relative (arc with a) coordinates of the end point of the arc.

9.4   Animations

You can change the properties of an SVG element using animation. As in a movie, this involves a continuous change in position, size, color, transparency, or other properties.

You can determine the start time and duration of an animation. Animations can run simultaneously or sequentially with respect to one or more SVG elements, and they can be triggered by certain events and repeated as often as required.

9.4.1 Procedure

The *svg_animation.htm* file contains some examples of animations. First, you'll see two rectangles with gray filling (see Figure 9.8). After a short time, an animation for the first rectangle begins. It moves downward, and then it moves back up to the starting position and becomes higher at the same time. Finally, it becomes almost transparent (see Figure 9.9).

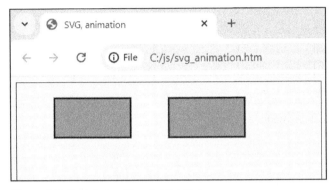

Figure 9.8 First Animation, Initial State

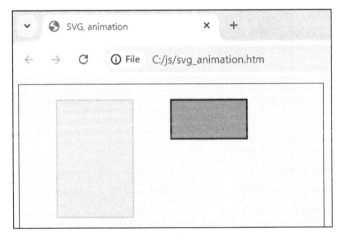

Figure 9.9 First Animation, Final State

A click on the second rectangle starts an animation for this element. It moves downward and turns white at the same time (see Figure 9.10), and then it moves back up to the start position and turns gray again at the same time. The entire process is repeated once.

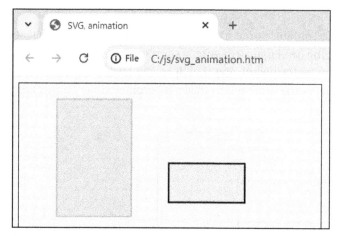

Figure 9.10 Second Animation, Intermediate State

9.4.2   Time Control

Here's the code for the time-controlled animations of the first rectangle:

```
...
<body>
    <svg xmlns="https://www.w3.org/2000/svg" width="400px"
            height="300px" style="border:1px solid black;">
        <g stroke="#000000" stroke-width="2" fill="#a0a0a0">
            <rect x="50" y="20" width="100" height="50">
                <animate attributeName="y"
                    begin="1" dur="1" from="20" to="120" fill="freeze" />
                <animate attributeName="height"
                    begin="2" dur="1" by="100" fill="freeze" />
                <animate attributeName="y"
                    begin="2" dur="1" by="-100" fill="freeze" />
                <animate attributeName="opacity"
                    begin="3" dur="1" from="1" to="0.2" fill="freeze" />
            </rect>
...
        </g>
    </svg>
</body></html>
```

Listing 9.9 svg_animation.htm File, Animations for First Rectangle

The two rectangles are in a group in which the properties for the fill color and the frame line are defined. Each of the four different animations is subordinate to the rectangle as child nodes.

You use the animate tag to create an animation for the parent element, and you use the attributeName attribute to define the property that you want to animate. In this case, these are the attributes y for the vertical position, height for the height, and opacity for the transparency.

I am only introducing the transparency property for SVG elements here because it's clearer in connection with an animation. An SVG element that lies behind a partially or completely transparent other SVG element is visible, and the permitted values for the associated opacity attribute are between 0.0 (= completely transparent) and the default value 1.0 (= completely opaque).

The begin and dur (short for *duration*) attributes define the start time and duration of a time-controlled animation. This enables you to write a *script* for the process. By default, the values are specified in terms of seconds, and in this example, each animation lasts one second. The start times are selected in such a way that the first animation takes place first, the second and third animations immediately follow it simultaneously, and the fourth animation immediately follows them.

You can also specify a time value can with an abbreviation for a specific unit. For example, h is possible for the hours unit (e.g., 2h), min for the minutes unit (e.g., 0.5min), s for the seconds unit (e.g., 3.5s), or ms for the milliseconds unit (e.g., 500ms).

You use the from and to attributes to determine the absolute start and end values for the animated property. To avoid a sudden change at the start of the animation, you should take the value that the property already has as the start value. In this example, the value for y is changed continuously from 20 to 120 during the first animation, and the value for opacity is changed continuously from 1.0 to 0.2.

You use the by attribute to determine the relative change in the animated property. In this example, the end value for height is 100 greater than the start value, while the end value for y in the third continuous animation is 100 less than the start value.

The fill attribute for the animation has nothing to do with a filling, and the freeze value ensures that the value gets frozen at the end of the animation. If you omit the attribute or set the default remove value, the start value for the property will be set again.

9.4.3 Event Control

Let's now look at the code for the event-controlled animations of the second rectangle:

```
...
        <rect id="re2" x="200" y="20" width="100" height="50">
          <animate attributeName="y" begin="re2.click" dur="2"
            repeatCount="2" values="20;120;20" />
          <animate attributeName="fill" begin="re2.click" dur="2"
```

```
                repeatCount="2" values="#a0a0a0;#ffffff;#a0a0a0" />
        </rect>
```

. . .

Listing 9.10 svg_animation.htm File, Animations for Second Rectangle

The rectangle has the id attribute, here with the re2 value. A unique ID is required to assign events to SVG elements. Each of the two different animations is subordinate to the rectangle as child nodes.

The begin attribute has the re2.click value for both animations, so a click on the rectangle with the corresponding ID starts both animations simultaneously. In addition to click, there are mouseover, mouseout, and other designations for events, as you have already seen in Chapter 4, but without the on prefix.

Both animations last 2 seconds. The repeatCount attribute defines the number of animation sequences, so a value of 2 doesn't mean that the animation will run two more times after it has finished but that it will run twice in total.

The values attribute is particularly useful if an animation runs from a start value via several intermediate values to a final value. In this case, the value of y is first changed continuously from 20 to 120 and then continuously from 120 to 20 again. The value for the fill color starts at gray, changes continuously to white, and then changes continuously to gray again. In both cases, the end value corresponds to the start value, so the fill attribute with the freeze value can be omitted.

9.5   Rotations

The rotation of an SVG element using the rotate statement belongs to the group of transformations. I am only introducing it here because it's clearer in the context of an animation.

In addition to rotation, there are other types of transformations that are not explained here. You can achieve them, with some effort, by changing a property:

- You can achieve the shift or translation using the translate statement by using other position specifications.

- You can achieve the scaling using the scale statement by using other size specifications.

- You can achieve the distortion using the skewX and skewY statements somewhat more elaborately by using other shape specifications.

Within the *svg_rotation.htm* file, you'll see three rectangles with a gray fill after loading the document (see Figure 9.11).

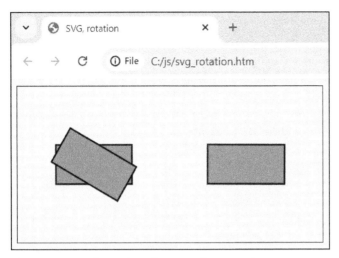

Figure 9.11 Fixed Rotation of Front Left Rectangle

The two rectangles on the left-hand side are on top of each other. The upper rectangle is rotated by 30 degrees in relation to the lower rectangle, and the pivot point (or more precisely, the point at which the axis of rotation pierces the screen) is the center of the rectangle. You can also define other pivot points (e.g., a corner of the rectangle or the zero point of the SVG graphic at the top left).

After a short time, an animation starts for the rectangle on the right-hand side. It rotates continuously from the start position at 0 degrees to the end position of 30 degrees (see Figure 9.12).

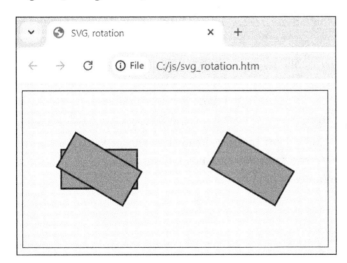

Figure 9.12 After Animated Rotation of Right Rectangle

Here's the code:

```
...
<body>
   <svg xmlns="https://www.w3.org/2000/svg" width="400px"
        height="200px" style="border:1px solid black;">
     <g stroke="#000000" stroke-width="2" fill="#a0a0a0">
       <rect x="50"  y="75" width="100" height="50" />
       <rect x="50"  y="75" width="100" height="50"
          transform="rotate(30, 100, 100)" />
       <rect x="250" y="75" width="100" height="50">
          <animateTransform attributeName="transform"
             type="rotate" begin="1" dur="2" from="0, 300, 100"
             to="30, 300, 100" fill="freeze" />
       </rect>
     </g>
   </svg>
</body></html>
```

Listing 9.11 svg_rotation.htm File

The two rectangles are in a group in which the properties for the fill and the frame line are defined.

The upper rectangle on the left-hand side has the transform attribute. The type of transformation is specified as the value for this attribute, such as translate, scale, skewX, skewY, or (as in this case) rotate, followed by values in parentheses. In a rotation, this is the value for the rotation angle in degrees, followed by the coordinates of the pivot point.

The animation is a subordinate child node of the rectangle on the right-hand side. You use the animateTransform tag to create an animation for a transformation of the parent element, and the attributeName attribute is given the transform value.

In the case of an animation for a rotation, the type attribute is given the rotate value. The effect of the begin, dur, and fill attributes is already known. The values of the from and to attributes are noted in the same way as above: first, the angle of rotation in degrees, and then, the coordinates of the pivot point.

9.6   SVG and JavaScript

With the help of JavaScript and the DOM you know from Chapter 5, you can access individual elements of SVG graphics, change them, or complement them.

Using the program in *svg_event.htm*, you can move a gray rectangle a little to the right after each click you make on the rectangle using JavaScript. You can also do this directly in SVG, but we'll show you how to use JavaScript to register events relating to an SVG element, read the values of its properties, and change them.

Here's the code:

```
... <head> ...
  <script>
    function change()
    {
      let xPos = parseInt(re.getAttribute("x"));
      xPos += 100;
      re.setAttribute("x", xPos);
    }
  </script>
</head>
<body>
  <svg xmlns="https://www.w3.org/2000/svg" width="400px"
      height="150px" style="border:1px solid black;">
    <rect id="re" x="50" y="20" width="100" height="50"
      fill="#a0a0a0" stroke="#000000" stroke-width="2" />
  </svg>
  <script>
    const re = document.getElementById("re");
    re.addEventListener("click", change);
  </script>
</body></html>
```

Listing 9.12 svg_event.htm File

The rectangle has an ID that, as with HTML elements, is required for the connection to JavaScript using the getElementById() method. In the change() function, the getAttribute() method returns the value of an attribute as a string, and the name of the attribute is specified in a string.

If the value is to be further processed as a number, it must first be converted using the parseInt() method. In this case, the numerical value is increased by 100, and this new value is then assigned to the property using the setAttribute() method. It expects two parameters: the name of the attribute as a string and the new value of the property.

9.7 Dynamic SVG Elements

You can use JavaScript not only to access existing SVG elements but also to create new SVG elements. This allows you to create dynamic SVG graphics with many similar elements.

9.7.1   Sequence of the Animation

The program in the *svg_create.htm* file generates a total of 36 rectangles. They are given random colors and are arranged in a 6 × 6 grid (see Figure 9.13).

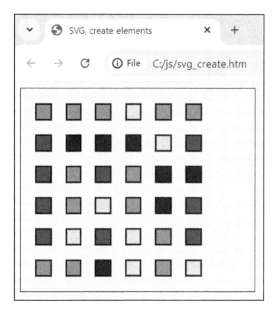

Figure 9.13 SVG Elements Created Using JavaScript

All rectangles then undergo similar animations: they are moved to the right, rotated by 90 degrees with a time delay (see Figure 9.14), and then given a different random color.

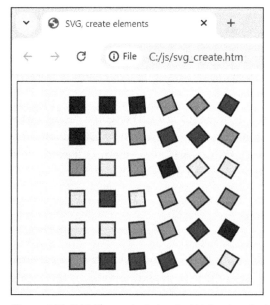

Figure 9.14 SVG Elements Animated Using JavaScript

9.7.2 Creating the Start State

The code for creating the start state is as follows:

```
... <head> ...
   <script>
      function create()
      {
         const svgCont = document.getElementById("idSVG");
         const ns = "http://www.w3.org/2000/svg";
         const colorArray = ["#ff0000", "#ffff00", "#00ff00", "#0000ff"];

         for(let i=0; i<6; i++)
         {
            for(let k=0; k<6; k++)
            {
               /* Add rectangle to SVG container */
               const re = document.createElementNS(ns, "rect");
               const xValue = 20 + 40 * i;
               re.setAttribute("x", xValue);
               const yValue = 20 + 40 * k;
               re.setAttribute("y", yValue);
               re.setAttribute("width", "20");
               re.setAttribute("height", "20");
               const fillValue = colorArray[Math.floor(Math.random() * 4)];
               re.setAttribute("fill", fillValue);
               re.setAttribute("stroke", "#000000");
               re.setAttribute("stroke-width", "2");
               svgCont.appendChild(re);
               ...
            }
         }
      }
   </script>
</head>
<body>
   <svg xmlns="https://www.w3.org/2000/svg" id="idSVG" width="310px"
      height="260px" style="border:1px solid black;" />
   <script>
      create();
   </script>
</body></html>
```

Listing 9.13 svg_create.htm File, Start State

The document body contains an SVG container with an ID that doesn't yet contain any elements. In this case, no end tag is required, but the start tag must be terminated with />.

Below this, the JavaScript function create() is called. This sequence ensures that the HTML document with the SVG container is loaded before the SVG elements are created.

In the create() function, some variables are set first. The svgCont variable references the SVG container, while the ns variable references the namespace required for the SVG elements. The colorArray field contains the four possible colors for selection using the random generator.

The arrangement of the 36 SVG elements within a 6 × 6 grid is made possible by means of a nested loop. To create a new SVG element, you use the createElementNS() method, with which you can create an element from a specific namespace. The method expects two parameters: the namespace and the type of element, which in this case is rect. The method has a problem with the specification https, therefore http is still used here.

You assign the attributes and their values to the newly created SVG element using the setAttribute() method. You determine the values of the x and y coordinates depending on the current row and the current column within the 6 × 6 grid, and you determine the value for the color using the random generator and the previously created field.

Once you have assigned all attributes, the newly created rectangle gets subordinated as a child node to the SVG container using the appendChild() method.

9.7.3   Creating Animations

Let's now take a look at the code for the animations. It's located within the inner loop and is therefore executed for one rectangle at a time:

```
...
/* Add animated translation to rectangle */
const animX = document.createElementNS(ns, "animate");
animX.setAttribute("attributeName", "x");
animX.setAttribute("begin", "1");
animX.setAttribute("dur", "1");
animX.setAttribute("by", "50");
animX.setAttribute("fill", "freeze");
re.appendChild(animX);

/* Add animated rotation to rectangle */
const animRot = document.createElementNS(ns, "animateTransform");
animRot.setAttribute("attributeName", "transform");
```

```
        animRot.setAttribute("type", "rotate");
        const beginValue = 3 + 0.2 * i;
        animRot.setAttribute("begin", beginValue);
        animRot.setAttribute("dur", "1");
        const fromValue = "0, " + (80+40*i) + ", " + (30+40*k);
        animRot.setAttribute("from", fromValue);
        const toValue = "90, " + (80+40*i) + ", " + (30+40*k);
        animRot.setAttribute("to", toValue);
        animRot.setAttribute("fill", "freeze");
        re.appendChild(animRot);

        /* Add animated color change to rectangle */
        const animFill = document.createElementNS(ns, "animate");
        animFill.setAttribute("attributeName", "fill");
        animFill.setAttribute("begin", "6");
        animFill.setAttribute("dur", "1");

        /* Color must change */
        let fillToValue;
        do
            fillToValue = colorArray[Math.floor(Math.random() * 4)];
        while(fillToValue == fillValue);
        animFill.setAttribute("to", fillToValue);

        animFill.setAttribute("fill", "freeze");
        re.appendChild(animFill);
...
```

Listing 9.14 svg_create.htm File, Animations

To create the animated translation, you create a new SVG element of the animate type using the createElementNS() method. After you assign the attributes and their values using the setAttribute() method, the newly created animation is subordinated as a child node to the respective rectangle using the appendChild() method.

You use the createElementNS(), setAttribute(), and appendChild() methods to create and assign the animated rotation. This time, the SVG element has the animateTransform type, and to ensure that the rotation is delayed, you set the value for the begin attribute individually, depending on the current column within the 6 × 6 grid. Please note the following for the values for the from and to attributes: the rotation takes place from the start state of 0 degrees to the end state of 90 degrees, and each rectangle rotates around its own center.

You also use the createElementNS(), setAttribute(), and appendChild() methods to create and assign the animated color change. This time, the SVG element has the animate type again, and you use a do-while loop to ensure that the new color value and the old color value are always different.

Note

You'll find the u_svg exercise in bonus chapter 1, section 1.19, in the downloadable materials for this book at *www.rheinwerk-computing.com/5875.*

Chapter 10

Three-Dimensional Graphics and Animations Using Three.js

You can use the Three.js library to display three-dimensional graphics and animations.

Three.js is a JavaScript library that you can use to display three-dimensional (3D) graphics in a browser, without additional programs. These 3D graphics can also contain animations.

Three.js is based on *Web Graphics Library* (WebGL), a programming interface for graphics that is a component of many modern browsers. The library is constantly being improved, partly thanks to an active community.

Currently (as of August 2024), Three.js is available in revision 167. The *three.min.js* file of revision 157 with the entire library has a size of 634 KB, and you can find it together with the sample files in the downloadable materials for this book. We access this file in the following examples, and it's located in the same directory as the sample files.

You can find any more recent versions at *https://threejs.org*, and you can use the **Download** link to download the *three.js-master.zip* file with a size of approximately 260 MB. The desired *three.min.js* file is located in the *build* subdirectory of the unpacked package. Occasionally, there are problems with the download, so if that happens to you, please use the mentioned file from the downloadable materials for this book.

This chapter covers the basics of Three.js: different three-dimensional geometries, different camera positions, and displacement, rotation, and animation of three-dimensional bodies. It does not cover advanced topics like different materials, types of light, and casting shadows.

10.1 First 3D Graphic

The individual elements of a 3D graphic are illustrated using the simple 3D geometry of a cube (see Figure 10.1).

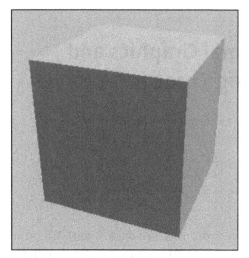

Figure 10.1 Cube as First 3D Object

We look at the cube with the camera from the top right. In this way, three faces of the cube are visible in different colors.

10.1.1   3D Coordinate System

To clarify the structure of the 3D graphic, we must first deal with the basics of a 3D coordinate system. Each point in 3D space can be uniquely identified using its x coordinate, y coordinate, and z coordinate.

The coordinates x=0, y=0, and z=0 denote the center of the system. This item is located in the center of the screen by default.

- The value of the x coordinate is positive to the right of the center and negative to the left of the center. The x axis of our system runs from left to right in the screen plane.

- Above the center, the value of the y coordinate is positive, and below the center, it's negative. The y axis of our system runs from bottom to top in the screen plane.

- The z axis of our system runs vertically through the screen plane, from a point behind the screen toward us as the viewer. The value of the z coordinate is positive in front of the screen and negative behind the screen.

In Figure 10.2, you can see the 3D coordinate system with the three axes. To illustrate the 3D space in the two-dimensional (2D) plane of the book, I choose a perspective representation, in which objects that are farther away from the viewer appear smaller. For the same reason, I use a perspective camera in my examples to display the 3D graphics in the 2D plane of the screen.

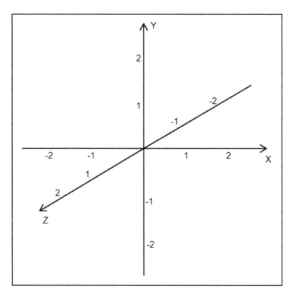

Figure 10.2 3D Coordinate System with X, Y, and Z Axes

10.1.2   Structure of the Program

The following is the code of the program in which the objects of the Three.js library are used. In the program code, I refer to these objects as *Three.js objects*, in contrast to the 3D objects, which can be seen on the screen.

The structure of my 3D graphics is always similar, so my extensive explanation of this first program will also help you learn about the other programs in this chapter.

```
... <head> ...
  <script src="three.min.js"></script>
</head>
<body id="idBody">
  <script>
    /* 3D object */
    const geometry = new THREE.BoxGeometry(200, 200, 200);
    const material = new THREE.MeshNormalMaterial();
    const object = new THREE.Mesh(geometry, material);

    /* Camera */
    const camera = new THREE.PerspectiveCamera(45,
        innerWidth/innerHeight, 1, 100000);
    camera.position.set(200, 250, 500);
    camera.lookAt(object.position);

    /* Drawing area */
    const drawing = new THREE.WebGLRenderer();
```

```
        drawing.setSize(innerWidth*0.8, innerHeight*0.8);
        drawing.setClearColor(0xc0c0c0);

        /* 3D scene */
        const scene = new THREE.Scene();
        scene.add(object);
        drawing.render(scene, camera);

        /* Add drawing to document */
        document.getElementById("idBody").appendChild(drawing.domElement);
</script>
</body></html>
```

Listing 10.1 th_grafic.htm File

At the beginning of the program, the library in the *three.min.js* file gets included.

10.1.3   3D Object with Geometry and Material

A 3D object has a shape, which is also known as a geometry. The geometry of a 3D object is made up of many individual triangular surfaces, and the Three.js library provides a range of ready-made, frequently used geometries, including the geometry of a sphere, a cuboid, and a cone. These finished geometries will save you a lot of assembly work.

You can create a box by using a Three.js object of the BoxGeometry type. The constructor expects at least three parameters to specify the width (in the x direction), the height (in the y direction) and the depth (in the z direction) of the cuboid. If you choose three identical values, you get the special case of a cube.

A 3D object has a surface material, and a Three.js object of the MeshNormalMaterial type appears in red from the x direction, green from the y direction, and blue from the z direction.

You can use a Three.js object of the Mesh type to create the 3D object in the desired geometry and with the desired surface material. No position is specified for the object, so its position is at the origin of the coordinate system.

10.1.4   Camera

There are different types of cameras. A Three.js object of the PerspectiveCamera type provides a camera that gives us a perspective view of the graphic scene. This camera stands on the top of a pyramid, which has a rectangular base.

In Figure 10.3, the camera is indicated by a thick dot on the right-hand side. The camera looks through the pyramid to the base, which is on the left-hand side. Everything outside the pyramid can't be seen in the graphic scene.

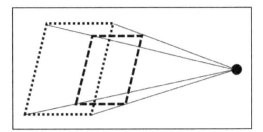

Figure 10.3 Perspective Camera with Layers

The constructor requires the following four parameters:

- fov = *field of view*: This is the viewing angle in degrees. A natural perspective is achieved at 45 degrees.
- aspect = *aspect ratio*: This is the ratio of the width to the height of the base of the rectangular pyramid. I recommended that you use the aspect ratio of the browser window.
- near = *near plane*: This is a (dashed) cross-section through the pyramid in Figure 10.3. Everything in front of this layer can't be seen in the graphic scene.
- far = *far plane*: This is the base of the pyramid (which is dotted in Figure 10.3). Everything behind this layer is also not visible in the graphic scene.

Using a Three.js object of the position type, you have access to the property of the position of a Three.js object. You can call or set the coordinates individually using the x, y, and z properties of the position object, or you can use the set() method to set all three coordinates. The x=200, y=250, and z=500 values selected here position the camera at the top right and at the front in relation to the origin of the coordinate system.

In the standard case, the viewing direction of the camera is exactly straight ahead, but you can use the lookAt() method to change the viewing direction so that the camera looks towards the origin.

10.1.5 Canvas and Scenes

A Three.js object of the WebGLRenderer type provides a drawing area, which is the canvas. The setSize() method is used to set the size of the canvas, and I use 80% of the width and 80% of the height of the browser window, using the values of the innerWidth and innerHeight properties of the window object.

The setClearColor() method uses a hexadecimal number sequence to determine the red, green, and blue components of the background color of the canvas. The default value is black.

A Three.js object of the Scene type creates a graphic scene in which 3D objects can be arranged, and the add() method adds the previously created 3D object (i.e., the cube) to the scene.

The render() method of the canvas is used to display the scene together with the camera within the canvas.

The domElement property of the canvas references the canvas itself. It's added to the document using the appendChild() method.

10.2   Moving the Camera

To illustrate the three-dimensional structure, we can move the camera in all three directions using the following program.

Here, we only show the part of the code that has been added compared to the program from Section 10.1:

```
...
<body id="idBody">
   <form>
   <script>
      ...
      let cx = 200, cy = 250, cz = 500;
      camera.position.set(cx, cy, cz);
      ...
   </script>

   <p>
   <input id="posPX" type="button" value="+X">
   <input id="posMX" type="button" value="-X">
   <input id="posPY" type="button" value="+Y">
   <input id="posMY" type="button" value="-Y">
   <input id="posPZ" type="button" value="+Z">
   <input id="posMZ" type="button" value="-Z"></p>
   </form>

   <script>
      document.getElementById("posPX").addEventListener
         ("click", function(){move(50,0,0)});
      document.getElementById("posMX").addEventListener
         ("click", function(){move(-50,0,0)});
      document.getElementById("posPY").addEventListener
         ("click", function(){move(0,50,0)});
      document.getElementById("posMY").addEventListener
```

```
        ("click", function(){move(0,-50,0)});
    document.getElementById("posPZ").addEventListener
        ("click", function(){move(0,0,50)});
    document.getElementById("posMZ").addEventListener
        ("click", function(){move(0,0,-50)});

    function move(dx, dy, dz)
    {
        cx += dx;
        cy += dy;
        cz += dz;
        camera.position.set(cx, cy, cz);
        camera.lookAt(object.position);
        drawing.render(scene, camera);
    }
    </script>
</body></html>
```

Listing 10.2 th_camera.htm File

The middle section of the document introduces the cx, cy, and cz variables, which specify the position of the camera.

In the lower part of the document, six buttons are created to implement a positive or negative movement of the camera in the *x*, *y*, and *z* directions (see Figure 10.4). Clicking the buttons calls the move() function, and three displacement values for the three directions are passed as parameters.

In the move() function in the lower part of the document, the three transferred move values are added to the current position values of the camera. The camera receives the newly determined position, and then the program makes sure that the camera looks at the cube again. Finally, the drawing must be rerendered.

As the absolute value of the camera's position values increases, the distance of the camera from the object increases as well so that the object becomes smaller and smaller from the viewer's perspective.

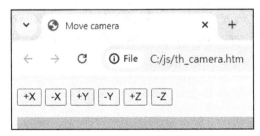

Figure 10.4 Buttons for Moving

10.3   Animation

To further illustrate the 3D structure, the cube from the first program in the section gets slowly rotated by a total of 90 degrees around the *y* axis, using an animation.

Here, we only show the part of the code that has been added compared to the program from Section 10.1:

```
...
<body id="idBody">
   <script>
      ...
      function rotate()
      {
         object.rotation.y += 0.2 * Math.PI / 180;
         drawing.render(scene, camera);
         if(object.rotation.y <= 90 * Math.PI / 180)
            requestAnimationFrame(rotate);
      }
      rotate();
   </script>
</body></html>
```

Listing 10.3 th_animation.htm File

You can use a Three.js object of the rotation type to access the rotation angle of a Three.js object. You can call or set this angle individually around the relevant axes using the x, y, or z properties of the rotation object, and you must specify the angle in radians. If you specify it in degrees, you must first multiply it by the factor $\pi / 180$.

The value of the angle of rotation around the *y* axis is increased by 0.2 degrees, which seems very small at first. The graphic scene is then redrawn using the render() method, and if the rotation angle around the *y* axis is less than or equal to 90 degrees, then the requestAnimationFrame() method of the window object gets called, again with a reference to the rotate() method. This ensures that the rotate() method will be called again. Then, the object is rotated again by 0.2 degrees, the graphic scene is drawn again, and so on. This is how you create the animation. By default, these actions take place sixty times per second, which results in a rotation of 12 degrees per second. The final value of 90 degrees is therefore reached after 7.5 seconds. If you remove the branch, the 3D object continues to rotate endlessly.

10.4   Various Shapes

In the downloadable materials for this book, you can find the *th_multiple.htm* program as a bonus. This program can be used to display a total of eight different shapes in grid view: a sphere, a cuboid, a cone, a tetrahedron, an octahedron, a capsule, a torus, and a torus knot (see Figure 10.5). The camera looks at the origin, which is in the center of the drawing. This is why the four shapes on the far left and far right are distorted in perspective.

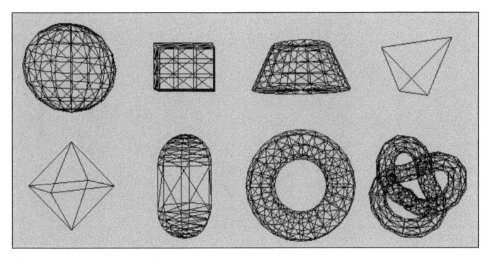

Figure 10.5 Eight Different Shapes

Note

The *th_change.htm* program is described in bonus chapter 4 in the downloadable materials for this book at *www.rheinwerk-computing.com/5875*. This program allows the eight different geometric shapes from Figure 10.5 to be displayed individually and with greater clarity. They can also be animatedly moved and rotated.

Chapter 11
jQuery

The browser-independent, standardized methods of the jQuery library have become indispensable for many websites.

jQuery is the most widely used JavaScript library, and it's used as the basis for many content management systems and web frameworks, such as Joomla and WordPress. jQuery provides convenient, browser-independent methods using CSS, Ajax, and animations, among other things. I have used the current (as of August 2024) version 3.7.1 for the examples in this chapter.

The *jquery-3.7.1.min.js* file with the entire library is only 86 KB in size. You can find it together with the sample files in the downloadable materials for this book.

If you want to download any more recent versions, you should visit *https://jquery.com*. You can find the file used here on the **Download** page via the *compressed production* term. In the following examples, I assume that the file is located in the same directory as the sample files.

11.1 Structure

Using an initial example, I will explain different ways of using jQuery. Figure 11.1 shows a file that contains a paragraph with text and three paragraphs with a hyperlink.

After clicking on the various hyperlinks, the content of the first paragraph changes due to the use of jQuery, as you can see in Figure 11.2, Figure 11.3, and Figure 11.4.

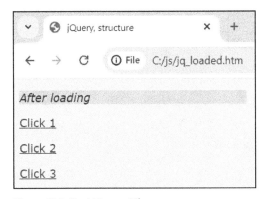

Figure 11.1 First jQuery File

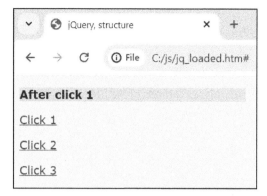

Figure 11.2 First Change

Figure 11.3 Second Change

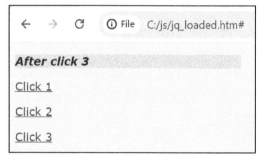

Figure 11.4 Third Change

Here's the code:

```
... <head> ...
   <script src="jquery-3.7.1.min.js"></script>
   <script>
     $(document).ready(function()
     {
```

```
        $("#idParagraph").html("<i>After loading</i>");
        $("#idLink1").click(function()
            { $("#idParagraph").html("<b>After click 1</b>"); });
    });
    </script>
</head>
<body>
    <p id="idParagraph" style="background-color:#e0e0e0;
        width:300px;">Hello</p>
    <p><a id="idLink1" href="#">Click 1</a></p>
    <p><a id="idLink2" href="#">Click 2</a></p>
    <p><a id="idLink3" href="#">Click 3</a></p>

    <script>
        $("#idLink2").click(function(){
            $("#idParagraph").html("After click 2"); });
        jQuery("#idLink3").click(function(){
            jQuery("#idParagraph").html("<b><i>After click 3</i></b>"); });
    </script>
</body></html>
```

Listing 11.1 jq_loaded.htm File

The document contains a paragraph with the sample text Hello. Below this, there are three paragraphs, each of which contains a hyperlink.

The ready() method is called in the upper JavaScript area, and it has a reference to a callback function as a parameter. Internally, the ready() method ensures that the callback function is only called after the file has been loaded with all elements in the browser. Otherwise, an element that doesn't yet exist could be accessed.

jQuery often uses anonymous callback functions as follows:

- The first statement in the function calls the html() method for the element with the idParagraph ID. It changes the text of an element including the HTML tags. Thus, once the document has been fully loaded, the first paragraph will display in italics.

- The second statement calls the click() method for the element with the idLink1 ID, and it also has a reference to an anonymous callback function as a parameter. Internally, the click() method ensures that the callback function is called after a click on the hyperlink. A text in bold then appears in the first paragraph.

At the end of the document, after you click on the second or third hyperlink, the click() method is called and the content of the first paragraph changes.

It's no longer necessary to call the ready() method here, as all elements of the document have already been loaded.

345

You can call a jQuery statement using both the $ function and the jQuery function. Both calls lead to the same result.

> **Note**
>
> A jQuery statement is executed for the element named in the selectors, which are often CSS selectors.
>
> Due to the many levels of parentheses in jQuery statements, I recommend that you use the compact notation I use here. In your own programs, please also note the frequently required quotation marks.

11.2   Selectors and Methods

You use selectors to select the element that is referenced by the jQuery code. We present the following selectors in this section:

- css(), to change the CSS properties
- html(), to change the text with HTML code
- text(), to change the text without HTML code

We use various other methods as well. In Figure 11.5, you can see four different div elements. You can change them by using the hyperlinks.

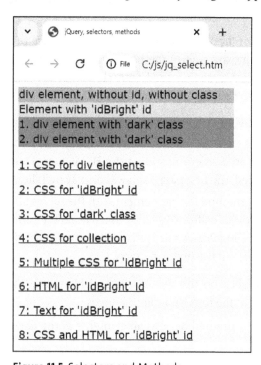

Figure 11.5 Selectors and Methods

Let's take a look at the code now, initially without the content of the ready() method:

```
... <head> ...
   <script src="jquery-3.7.1.min.js"></script>
   <script>
      $(document).ready(function() { ... });
   </script>
   <style>
      div        {width:250px; height:20px; background-color:#c0c0c0;}
      #idBright {background-color:#e0e0e0;}
      .dark      {background-color:#a0a0a0;}
   <style>
</head>
<body>
   <div>div element, without id, without class</div>
   <div id="idBright">Element with 'idBright' id</div>
   <div class="dark">1. div element with 'dark' class</div>
   <div class="dark">2. div element with 'dark' class</div>

   <p><a id="idLink1" href="#"> 1: CSS for div elements</a></p>
   <p><a id="idLink2" href="#"> 2: CSS for 'idBright' id</a></p>
   <p><a id="idLink3" href="#"> 3: CSS for 'dark' class</a></p>
   <p><a id="idLink4" href="#"> 4: CSS for collection</a></p>
   <p><a id="idLink5" href="#"> 5: Multiple CSS for 'idBright' id</a></p>
   <p><a id="idLink6" href="#"> 6: HTML for 'idBright' id</a></p>
   <p><a id="idLink7" href="#"> 7: Text for 'idBright' id</a></p>
   <p><a id="idLink8" href="#"> 8: CSS and HTML for 'idBright' id</a></p>
</body></html>
```

Listing 11.2 jq_select.htm File, without ready() Method

I explain the content of the ready() method next. The following are general settings for all div elements:

- They have a size of 250 × 20 pixels and are medium gray.
- The element with the idBright ID is light gray.
- All elements of the dark CSS class are dark gray.

Here are the four div elements:

- The first element has no ID and is not assigned to any CSS class.
- The second element has the idBright ID.
- The third and fourth elements have the CSS dark class assigned to them.

This is followed by eight paragraphs, each of which contains a link. The links have the IDs idLink1 through idLink8.

Now, here's the content of the ready() method:

```
$("#idLink1").click(function(){
    $("div").css({"width":"350px"}); });
$("#idLink2").click(function(){
    $("#idBright").css({"width":"400px"}); });
$("#idLink3").click(function(){
    $(".dark").css({"width":"450px"}); });
$("#idLink4").click(function(){
    $("#idBright, .dark").css({"width":"500px"}); });
$("#idLink5").click(function(){
    $("#idBright").css({"background-color":"#f0f0f0",
                        "width":"550px"}); });
$("#idLink6").click(function(){
    $("#idBright").html("<b>HTML new</b>"); });
$("#idLink7").click(function(){
    $("#idBright").text("Text new"); });
 $("#idLink8").click(function(){
    $("#idBright").css({"width":"+=20px"}).html("CSS and HTML new");});
```

Listing 11.3 jq_select.htm File, Content of ready() Method

After you click the first hyperlink, the css() method gets executed for all div elements and changes the CSS properties. Here, the width of the elements is set to 350 pixels, and you specify a property-value pair in JSON format. The second hyperlink changes the element with the idBright ID, and the third hyperlink changes all elements with the dark CSS class.

You can combine multiple selectors in one collection. The fourth hyperlink changes the element with the idBright ID and all elements with the dark CSS class.

In JSON format, multiple property-value pairs are separated by commas. The fifth hyperlink changes the background color and the width for the element with the idBright ID.

The sixth hyperlink calls the html() method, which changes the text including the HTML code.

The seventh hyperlink calls the text() method, which changes the text without taking the HTML code into account. Any HTML elements accidentally included would also be output as text.

You can use concatenation to execute multiple methods for one selector, and you can also change a CSS value in relation to the original value by using the += and -= operators. The eighth hyperlink executes the css() and html() methods for the element with the idBright ID one after the other, and in the css() method, the width is increased by 20 pixels per click.

11.3   Events

In addition to the click event, you can use other events to start jQuery code.

In Figure 11.6, you can see a div element with a series of hyperlinks below it. Triggering an event on a link leads to an animated expansion of the element, and you can use the animate() method to create an animation. You can find out more about animations in Section 11.4.

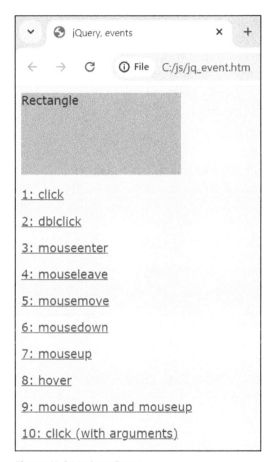

Figure 11.6 Various Events

Now, let's take a look at the code, initially without the content of the ready() method:

```
... <head> ...
   <script src="jquery-3.7.1.min.js"></script>
   <script>
      $(document).ready(function() { ... });
   </script>
</head>
```

```
<body>
    <div id="idRect" style="width:200px; height:100px;
        background-color:#c0c0c0;">Rectangle</div>
    <p><a id="idLink1" href="#"> 1: click</a></p>
    <p><a id="idLink2" href="#"> 2: dblclick</a></p>
    <p><a id="idLink3" href="#"> 3: mouseenter</a></p>
    <p><a id="idLink4" href="#"> 4: mouseleave</a></p>
    <p><a id="idLink5" href="#"> 5: mousemove</a></p>
    <p><a id="idLink6" href="#"> 6: mousedown</a></p>
    <p><a id="idLink7" href="#"> 7: mouseup</a></p>
    <p><a id="idLink8" href="#"> 8: hover</a></p>
    <p><a id="idLink9" href="#"> 9: mousedown and mouseup</a></p>
    <p><a id="idLink10" href="#">10: click (with arguments)</a></p>
</body></html>
```

Listing 11.4 jq_event.htm File, without ready() Method

The div element has the idRect ID, a size of 200 × 100 pixels, and a gray color.

Now, here's the content of the ready() method:

```
$("#idLink1").click(function(){
    $("#idRect").animate({"width":"+=20px"}); });
$("#idLink2").dblclick(function(){
    $("#idRect").animate({"width":"+=20px"}); });
$("#idLink3").mouseenter(function(){
    $("#idRect").animate({"width":"+=20px"}); });
$("#idLink4").mouseleave(function(){
    $("#idRect").animate({"width":"+=20px"}); });
$("#idLink5").mousemove(function(){
    $("#idRect").animate({"width":"+=20px"}); });
$("#idLink6").mousedown(function(){
    $("#idRect").animate({"width":"+=20px"}); });
$("#idLink7").mouseup(function(){
    $("#idRect").animate({"width":"+=20px"}); });
$("#idLink8").hover(function(){
    $("#idRect").animate({"width":"+=20px"}); });
$("#idLink9").bind("mousedown mouseup", function(){
    $("#idRect").animate({"width":"+=20px"}); });
$("#idLink10").click(function(e){
    $("#idRect").html("Event: " + e.type
        + "<br>Position X: " + e.pageX + " , Y: " + e.pageY
        + "<br>Time: " + Date(e.timeStamp)); });
```

Listing 11.5 jq_event.htm File, Content of ready() Method

Simply click on the first hyperlink to call the click() method. The element becomes 20 pixels wider, as with almost all subsequent events. Double-click on the second hyperlink to call the dblclick() method.

The third hyperlink gets triggered as soon as the mouse cursor enters the area above the hyperlink (i.e., by the mouseenter event). This calls the mouseenter() method, and the same thing happens after the mouse cursor leaves the area above the fourth hyperlink (the name of the event and method is mouseleave) and hovers over the fifth hyperlink (the name of the event and method is mousemove).

Pressing down a mouse button on the sixth hyperlink triggers the mousedown event, and releasing the mouse button on the seventh hyperlink triggers the mouseup event. These actions call the mousedown() and mouseup() methods, respectively.

The hover() method combines the mouseenter and mouseleave events, which are triggered as soon as the mouse cursor enters the area above the eighth hyperlink or leaves it (respectively).

You use the jQuery bind() method to bind events to methods. In fact, the other methods in this section are actually specializations of the bind() method in abbreviated form. You use the ninth hyperlink to bind the mousedown and mouseup events, so you trigger the animation by both pressing down and releasing a mouse button.

Information about each of the events is provided in an event object. With jQuery, this event object is standardized for all browsers, and you can access it via a reference that you pass to the method as a parameter. When you click on the tenth hyperlink, some information gets displayed: in this case, the type, location, and time of the event, using the type, pageX, pageY, and timeStamp properties. The timeStamp is specified in milliseconds, and you can convert it using the Date() jQuery method.

11.4   Animations

This section describes various options for animating elements. For this purpose, you use the animate() method, which (as in a movie) creates the impression of an even sequence of individual images.

In Figure 11.7, you can see a positioned div element. You can use the first eleven hyperlinks to start different animations for this element. For clarification, you should restore the initial state after loading the page after each animation. You can do this by using the twelfth hyperlink.

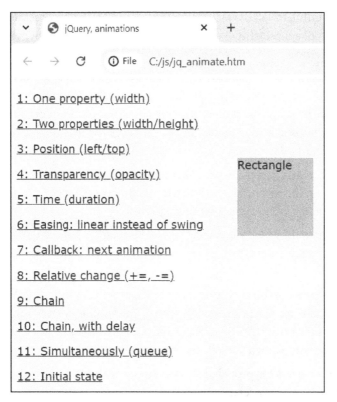

Figure 11.7 Animations

Now, let's take a look at the code, initially without the content of the ready() method:

```
... <head> ...
   <script src="jquery-3.7.1.min.js"></script>
   <script>
      $(document).ready(function() { ... });
   </script>
</head>
<body>
   <div id="idRect" style="position:absolute; width:100px; height:100px;
      left:300px; top:100px; background-color:#c0c0c0;">Rectangle</div>
   <p><a id="idLink1" href="#"> 1: One property (width)</a></p>
   <p><a id="idLink2" href="#"> 2: Two properties (width/height)</a></p>
   <p><a id="idLink3" href="#"> 3: Position (left/top)</a></p>
   <p><a id="idLink4" href="#"> 4: Transparency (opacity)</a></p>
   <p><a id="idLink5" href="#"> 5: Time (duration)</a></p>
   <p><a id="idLink6" href="#"> 6: Easing: linear instead of swing</a></p>
   <p><a id="idLink7" href="#"> 7: Callback: next animation</a></p>
   <p><a id="idLink8" href="#"> 8: Relative change (+=, -=)</a></p>
   <p><a id="idLink9" href="#"> 9: Chain</a></p>
```

```
    <p><a id="idLink10" href="#">10: Chain, with delay</a></p>
    <p><a id="idLink11" href="#">11: Simultaneously (queue)</a></p>
    <p><a id="idLink12" href="#">12: Initial state</a></p>
</body></html>
```

Listing 11.6 jq_animate.htm File, without ready() Method

The div element has the idRect ID, an initial size of 100 × 100 pixels, an initial position of 300 × 100 pixels, and a gray color. Positioning is only necessary when the position gets animated.

Now, here's the content of the ready() method:

```
$("#idLink1").click(function(){
    $("#idRect").animate({"width":"200px"}); });
$("#idLink2").click(function(){
    $("#idRect").animate({"width":"200px", "height":"50px"}); });
$("#idLink3").click(function(){
    $("#idRect").animate({"left":"400px", "top":"200px"}); });
$("#idLink4").click(function(){
    $("#idRect").animate({"opacity":"0.5"}); });
$("#idLink5").click(function(){
    $("#idRect").animate({"width":"200px"}, {"duration":2000});});
$("#idLink6").click(function(){
    $("#idRect").animate({"left":"400px"},
    {"duration":2000, "easing":"linear"}); });
$("#idLink7").click(function(){
    $("#idRect").animate({"left":"400px"},
        function(){$("#idRect").animate({"left":"300px"}) }); });
 $("#idLink8").click(function(){
    $("#idRect").animate({"left":"+=100px", "opacity":"-=0.3"});});
$("#idLink9").click(function(){
    $("#idRect").animate({"left":"+=100px"})
        .animate({"left":"-=100px"}); });
$("#idLink10").click(function(){
    $("#idRect").animate({"left":"+=100px"})
        .delay(1000).animate({"left":"-=100px"}); });
$("#idLink11").click(function(){
    $("#idRect").animate({"width":"200px"}, {"duration":1000})
        .animate({"height":"50px"}, {"duration":2000, "queue":false}); });
$("#idLink12").click(function(){
    $("#idRect").animate({"width":"100px", "height":"100px",
        "left":"300px", "top":"100px", "opacity":1.0}); });
```

Listing 11.7 jq_animate.htm File, Content of ready() Method

The first hyperlink animates the width up to the target value of 200 pixels. You can change multiple properties at the same time. The second hyperlink animates the width up to the target value of 200 pixels and the height up to the target value of 50 pixels.

You can achieve animated movement by changing the property values for left and top. The third hyperlink moves the rectangle to the target point of 400 pixels/200 pixels. You can use the fourth hyperlink to change the transparency to the value of 0.5 via the opacity property (see also Chapter 8, Section 8.2.4).

An animation takes 0.4 seconds (i.e., 400 milliseconds) without further specification. You can set the duration of the animation in the second parameter of the animate() method by using the duration property. You write the value in milliseconds, without quotation marks. The fifth hyperlink changes the width to the target value of 200 pixels within 2 seconds.

The easing property characterizes the timing of an animation. The default swing value means that the animation accelerates at the beginning, continues at a steady speed, and slows down at the end. This creates the impression of a natural process.

If you enter the linear value for the easing property in the second parameter of the animate() method, the animation runs at a constant speed, which doesn't look that natural. Easing plug-ins, which you can find on the internet, provide further options for easing functions. The sixth hyperlink is used to move the element linearly to the target value within 2 seconds.

You can also pass a reference to a callback function as an additional parameter, which is executed after the end of the animation. The seventh hyperlink moves the element to the target value of 400 pixels, and the element is then moved to the target value of 300 pixels.

So far, we have specified an absolute target value for the animated property, but you can also make relative changes using the += and -= operators. The eighth hyperlink moves the element by 100 pixels to the right each time it's clicked and changes the transparency by 0.3, based on the current values. A value above 1.0 or below 0.0 doesn't make sense for the opacity property, but it doesn't lead to an error.

Method calls can be concatenated. The ninth hyperlink is used to move the element by 100 pixels to the right and then 100 pixels to the left.

You can add a delay within an animation using the delay() method. The tenth hyperlink leads to the same movement as the ninth hyperlink, but there's a one-second wait between the two partial animations.

In a chain, the individual parts of an animation run one after the other by default, but you can use the queue parameter to ensure that they take place at the same time. The eleventh hyperlink animates the width to the target value of 200 pixels within one second. The height is animated to the target value of 50 pixels, but within 2 seconds and at the same time. This is achieved with the Boolean value false (without quotation marks) for the queue property. The default value is true.

The twelfth hyperlink is used to restore the initial state of a total of five properties: width, height, left, top, and opacity.

The events are buffered. If you click on a hyperlink before a running animation has finished, the corresponding action will be executed afterward.

The following methods don't provide any additional options, but they can be used as a shortcut:

- The slideDown(), slideUp(), and slideToggle() methods for changing the height property
- The fadeIn(), fadeOut(), fadeToggle(), and fadeTo() methods for changing the opacity property
- The show(), hide(), and toggle() methods for simultaneously changing the width, height, and opacity properties

11.5   Example: Sinusoidal Movement

So far, only jQuery methods have been called in the individual functions, but we shouldn't forget that the rest of JavaScript is also available to us. The following example shows a combination of jQuery and the rest of JavaScript. After clicking on the small square on the far left, it moves along a sine curve. In addition, there are some auxiliary lines (see Figure 11.8).

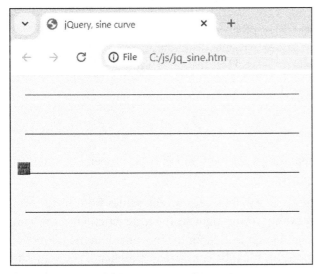

Figure 11.8 Animation of Sine Curve, Starting Point

The curve consists of individual line sections. Each line segment is generated by a jQuery animation, and a for loop ensures that the animations are sequenced. The finer the curve that is broken down, the smoother the curve progression, but you can already see a fairly homogeneous progression when it's divided into sections of 10 degrees each.

Here's the code:

```
... <head> ...
   <script src="jquery-3.7.1.min.js"></script>
   <script>
     function sine()
     {
        for(let angle = 10; angle <= 360; angle += 10)
        {
           const pLeft = (10 + angle) + "px";
           const pTop = (110 - Math.sin(angle / 180 * Math.PI) * 100) + "px";
           $("#idBlock").animate({"left":pLeft, "top":pTop},
              {"duration":"100", "easing":"linear"});
        }
     }

     $(document).ready(function() { $("#idLink").click(sine); });
   </script>
</head>
<body>
   <div id="idBlock" style="position:absolute; left:10px; top:110px;">
   <a id="idLink" href="#"><img src="block.gif" alt="Block"></a></div>
   <script>
      for(let pTop = 10; pTop < 211; pTop += 50)
         document.write("<div style='position:absolute; left:20px; top:"
            + pTop + "px;'><img src='im_line.jpg' alt='Line'></div>");
   </script>
</body></html>
```

Listing 11.8 jq_sine.htm File

The auxiliary lines are generated using a for loop, and the pTop property assumes values of 10, 60, 110, 160, and 210 pixels.

Within the sine() function, the angle variable takes on the values from 10 to 360 in steps of 10. The image moves evenly to the right in the x direction, and the sin() method of the Math object from JavaScript is used for the movement in the y direction. It expects the angle in radians, so this must be converted beforehand. The values for the target point calculated in this way are saved in the pLeft and pTop variables, and these variables can in turn be used as target values for the left and top properties as parameters of the animate() method.

11.6   jQuery and Ajax

Ajax allows you to reload document parts, as we described in Chapter 7. There are a number of methods in jQuery that use Ajax technology internally, and you can work independently of browser and version, as you are used to with jQuery.

In this example, you use the load() and post() methods to load the content from text files, HTML files, PHP programs, and XML files into the current document without having to rebuild the rest of the page.

You can see some examples in the following program. The start status of the page is shown in Figure 11.9, and you'll need to load the page via a web server (see also Chapter 7).

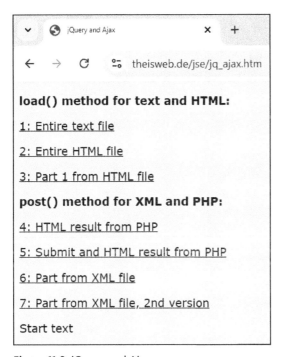

Figure 11.9 jQuery and Ajax

Here's the code, initially without the content of the second JavaScript container:

```
<!DOCTYPE html><html lang="en-us">
<head>
   <meta charset="utf-8">
   <title>jQuery and Ajax</title>
   <link rel="stylesheet" href="js5.css">
   <script src="jquery-3.7.1.min.js"></script>
   <script> ... </script>
</head>
```

```
<body>
    <p><b>load() method for text and HTML:</b></p>
    <p><a id="idLink1" href="#"> 1: Entire text file</a></p>
    <p><a id="idLink2" href="#"> 2: Entire HTML file</a></p>
    <p><a id="idLink3" href="#"> 3: Part 1 from HTML file</a></p>

    <p><b>post() method for XML and PHP:</b></p>
    <p><a id="idLink4" href="#"> 4: HTML result from PHP</a></p>
    <p><a id="idLink5" href="#"> 5: Submit and HTML result from PHP</a></p>
    <p><a id="idLink6" href="#"> 6: Part from XML file</a></p>
    <p><a id="idLink7" href="#"> 7: Part from XML file, 2nd version</a></p>

    <p id="idOutput">Start text</p>
</body></html>
```

Listing 11.9 jq_ajax.htm File, without Second JavaScript Container

The loaded content is displayed in the bottom paragraph with the idOutput ID.
Here's the content of the second JavaScript container:

```
function xmlData(result)
{
    const a = result.getElementsByTagName
        ("nodeA")[0].firstChild.nodeValue;
    const b = result.getElementsByTagName
        ("nodeB")[0].getAttribute("attributeA");
    $("#idOutput").html(a + ", " + b);
}

function xmlOutput()
{
    $.post("jq_ajax_test.xml",
        function(result) { xmlData(result); });
}

$(document).ready(function()
{
    $("#idLink1").click(function() {
        $("#idOutput").load("jq_ajax_test.txt"); });
    $("#idLink2").click(function() {
        $("#idOutput").load("jq_ajax_test.htm"); });
    $("#idLink3").click(function() {
        $("#idOutput").load("jq_ajax_test.htm #t1"); });
```

```
$("#idLink4").click(function() {
    $.post("jq_ajax_test.php", function(result) {
        $("#idOutput").html(result); }); });
$("#idLink5").click(function() {
    $.post("jq_ajax_test_data.php", {number1:12.2, number2:25.5},
        function(result) {$("#idOutput").html(result);}); });
$("#idLink6").click(function() {
    $.post("jq_ajax_test.xml", function(result) {
        $("#idOutput").html(result.getElementsByTagName("nodeA")[0]
        .firstChild.nodeValue + ", " + result.getElementsByTagName(
        "nodeB")[0].getAttribute("attributeA")); }); });
$("#idLink7").click( xmlOutput );
});
```

Listing 11.10 jq_ajax.htm File, Content of the ready() Method

The first hyperlink loads the entire text from the *jq_ajax_test.txt* text file into the paragraph using the load() method.

```
This is the text from the text file
```

Listing 11.11 jq_ajax_test.txt File

The second hyperlink loads the entire content of the HTML file *jq_ajax_test.htm* with the tags into the paragraph.

The third hyperlink only loads the content of the element with the t1 ID from the HTML file into the paragraph, also including the tags. Pay attention to the separating space between the file name and the hash character of the ID in the parameter of the load() method.

```
...
<body>
    <p><b>Text in HTML file</b></p>
    <p id="t1"><i>Part 1 in HTML file</i></p>
    <p id="t2"><i>Part 2 in HTML file</i></p>
</body></html>
```

Listing 11.12 jq_ajax_test.htm File

The fourth hyperlink and the post() jQuery method are used to call the PHP program in the *jq_ajax_test.php* file. The parameter of the callback function (here, result) then contains the output of the PHP program, and it becomes the content of the paragraph using the html() method.

```php
<?php
   header("Content-type: text/html; charset=utf-8");
   echo "<b>Data</b> from PHP file";
?>
```

Listing 11.13 jq_ajax_test.php File

The fifth hyperlink calls the PHP program *jq_ajax_test_data.php* using the `post()` jQuery method. Property-value pairs are sent, separated by colons. In this example, the sum of the two values is determined and returned in the PHP program, and this return becomes the content of the paragraph.

```php
<?php
   header("Content-type: text/html; charset=utf-8");
   echo "<b>Total</b>: "
      . (doubleval($_POST["number1"]) + doubleval($_POST["number2"]));
?>
```

Listing 11.14 jq_ajax_test_data.php File

The sixth hyperlink calls the *jq_ajax_test.xml* file, as well as the `post()` jQuery method. The value of the `nodeA` node and the value of the `attributeA` attribute of the `nodeB` node are determined and added to the content of the paragraph.

```xml
<?xml version="1.0" encoding="UTF-8"?>
<root>
  <nodeA>Node A value</nodeA>
  <nodeB attributeA = "Attribute A from node B"
      attributeB = "Attribute B from node B">
      Node B value</nodeB>
</root>
```

Listing 11.15 jq_ajax_test.xml File

The nesting of calls and anonymous functions in the sixth hyperlink is confusing, so I have added a seventh hyperlink that leads to the same result. Two named functions and partial results are used, and this makes it easier to understand the entire process.

Chapter 12
Mobile Apps Using Onsen UI

You can create user interfaces for mobile devices using JavaScript and the Onsen UI library.

Onsen UI is a framework that provides components for the development of mobile apps. These Onsen UI apps have user interfaces that are particularly suitable for mobile devices such as smartphones and tablets. The apps can also be easily linked with other technologies (e.g., JavaScript programs).

The current version (as of August 2024) is 2.12.8. You can download the required CSS and JavaScript files from *https://onsen.io*, but usually, pages are accessed online from mobile devices. I therefore use the online versions of the required files in my sample programs so that I always use the latest version.

The programs in this chapter were tested on a smartphone, and the images were created in the Google Chrome browser on the smartphone. You can also use a modern browser on a standard PC to test most of the program functions during development.

The Onsen UI framework uses its own style sheets, so I do not include the *js5.css* file (which ensures a uniform display using style sheets in most of the examples in this book) in the files in this chapter.

12.1 Structure of a Page

First, I want to describe the basic structure of a page, and later, I'll show you lists and tables that you can use to design a page.

12.1.1 First Page

The first page consists of a navigation bar and a content area (see Figure 12.1). The navigation bar can contain a page title and central operating elements, and the content area contains the actual content of the document: a list with one entry, namely, a button.

After you press the button, an informational message appears (see Figure 12.2).

Figure 12.1 First Page

The program reads as follows:

```
<!DOCTYPE html><html lang="en-us">
<head>
  <meta charset="utf-8">
  <title>Onsen UI, first page</title>
  <link rel="stylesheet" href="https://unpkg.com/onsenui/css/onsenui.css">
  <link rel="stylesheet"
      href="https://unpkg.com/onsenui/css/onsen-css-components.min.css">
  <script src="https://unpkg.com/onsenui/js/onsenui.min.js"></script>
</head>
<body>
<ons-page>
  <ons-toolbar>
    <div class="left" style="padding-left:10px;">Links</div>
    <div class="center">Center</div>
    <div class="right" style="padding-right:10px;">Right</div>
  </ons-toolbar>
  <ons-list>
    <ons-list-item>
      <ons-button id="idHello">Hello</ons-button>
    </ons-list-item>
  </ons-list>
</ons-page>

<script>
  document.getElementById("idHello").addEventListener("click",
      function(){ons.notification.alert("Hello");});
</script>
</body></html>
```

Listing 12.1 onsen_page.htm File

The header of the file contains the links to the online versions of the required CSS and JS files I mentioned at the beginning of this chapter. This looks the same in the other files, so the header of the file is no longer shown there.

Onsen UI components are integrated into the document like HTML containers. A component of the ons-page type forms the basic element of a page, which extends across the entire screen and serves as a container for other elements.

A page can contain a navigation bar using a component of the ons-toolbar type. The navigation bar is divided into three areas using div containers and the following CSS classes from Onsen UI:

- left, for elements on the left edge
- center, for elements that are slightly indented
- right, for elements on the right edge

In the left and right areas, the padding CSS property was used to create small gaps before the left and right edges.

You can use components of the ons-list and ons-list-item types to create a list or a list entry. More details on lists will follow in Section 12.1.2.

You can display buttons in the content area using a component of the ons-button type. You create the connection to JavaScript in the usual way, via the id attribute and the getElementById() and addEventListener() methods.

The component of the ons.notification type provides methods with various standard dialogs (see also Section 12.2.2), and the ons.notification.alert() method generates an information window with a message (see Figure 12.2).

Figure 12.2 Link to JavaScript

12.1.2 List of Elements

Lists are often used to display multiple elements in a clear way. You can assign a title to a list, group together related elements of a list, and divide each element of a list into three areas. You can also link list elements to actions.

In Figure 12.3, you can see a subdivided list.

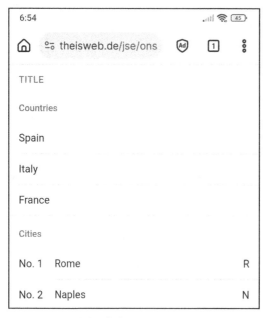

Figure 12.3 Subdivided List

Here's the associated program:

```
...
<body>
<ons-page>
  <ons-list>

    <ons-list-title>Title</ons-list-title>

    <ons-list-header>Countries</ons-list-header>

    <ons-list-item tappable id="idSpain"> Spain </ons-list-item>
    <ons-list-item tappable id="idItaly"> Italy </ons-list-item>
    <ons-list-item tappable id="idFrance"> France </ons-list-item>

    <ons-list-header>Cities</ons-list-header>

    <ons-list-item>
      <div class="left">No. 1</div>
      <div class="center">Rome</div>
      <div class="right">R</div>
    </ons-list-item>
```

```
    <ons-list-item>
       <div class="left">No. 2</div>
       <div class="center">Naples</div>
       <div class="right">N</div>
    </ons-list-item>

  </ons-list>
</ons-page>

<script>
   document.getElementById("idSpain").addEventListener("click",
      function(){ons.notification.alert("Madrid");});
   document.getElementById("idItaly").addEventListener("click",
      function(){ons.notification.alert("Rome");});
   document.getElementById("idFrance").addEventListener("click",
      function(){ons.notification.alert("Paris");});
</script>
</body></html>
```

Listing 12.2 onsen_list.htm File

A list consists of at least one component of the ons-list type for the entire list and of entries that are created using components of the ons-list-item type. A list can have an overall title within a component of the ons-list-title type, and you use components of the ons-list-header type to group elements and as group titles. You divide a single element into three areas as in a navigation bar using the left, center, and right classes.

Tapping a list item corresponds to an event that you can use to trigger actions (see Figure 12.4). The tappable attribute ensures a visible reaction of the list item itself.

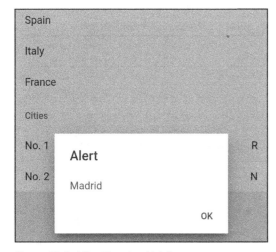

Figure 12.4 After Tapping List Item

12.1.3   Table with Items

Another option for displaying multiple elements is provided by components of the `ons-row` type and subordinate components of the `ons-col` type. This allows you to create individual *rows* in a similar way to a table. Within each row, you can create individual cells by using *columns*. In contrast to an HTML table, the rows are independent of each other, and you can set the number and width of the cells individually within each row. In Figure 12.5, you can see a table with three rows.

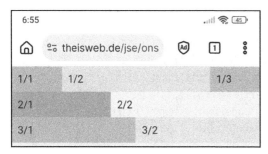

Figure 12.5 Table

Here's the associated program:

```
...
<body>
<ons-page>

  <ons-row>
    <ons-col width="80px"
      style="background-color:#c0c0c0; padding:10px;">1/1</ons-col>
    <ons-col style="background-color:#e0e0e0; padding:10px;">1/2</ons-col>
    <ons-col width="80px"
      style="background-color:#c0c0c0; padding:10px;">1/3</ons-col>
  </ons-row>

  <ons-row>
    <ons-col width="40%"
      style="background-color:#b0b0b0; padding:10px;">2/1</ons-col>
    <ons-col width="60%"
      style="background-color:#f0f0f0; padding:10px;">2/2</ons-col>
  </ons-row>

  <ons-row>
    <ons-col style="background-color:#c0c0c0; padding:10px;">3/1</ons-col>
    <ons-col style="background-color:#e0e0e0; padding:10px;">3/2</ons-col>
  </ons-row>
```

```
</ons-page>
</body></html>
```

Listing 12.3 onsen_table.htm File

The cells here have different background colors for a clearer display. Within a cell, you can use the padding CSS property to create a space between the content and the edge of the cell.

In the first row, there are two cells on the left and right, each with a fixed width of 80 pixels. The third cell in the middle takes up the remaining space in the row. The second row contains two cells that take up 40% and 60% of the width of the row. There are also two cells in the third row, and as their width is not specified, each of them takes up half the width of the row.

12.2   Elements within a Page

In this section, you'll get to know various elements that enable interaction. These include icons, dialog boxes, input fields, and selection fields.

12.2.1   Icons and Floating Action Buttons

You can integrate numerous icons as easily understandable symbols or pictograms as information or prompts for interaction. You create such icons as vector graphics like the characters of a font. This means that you can easily change the icons' size, color, and other properties.

A *floating action button* (FAB) is a circular button that typically triggers the most important action in an Android app. In Figure 12.6, you can see some examples from two extensive icon collections, while Figure 12.7 shows a FAB.

Figure 12.6 Various Icons

Figure 12.7 FAB at Bottom Edge of Screen

Here's the related program:

```
...
<body>
<ons-page>
  <ons-list>

    <ons-list-item>
      <ons-icon icon="md-face"></ons-icon>
      <ons-icon icon="md-home"></ons-icon>
      <ons-icon icon="md-zoom-in"></ons-icon>
    </ons-list-item>

    <ons-list-item>
      <ons-icon icon="fa-file"></ons-icon>
      <ons-icon icon="fa-cog"></ons-icon>
      <ons-icon icon="fa-car"></ons-icon>
    </ons-list-item>

    <ons-list-item>
      <ons-icon style="color:#ff0000;" icon="md-home"></ons-icon>
      <ons-icon style="color:#00ff00;" icon="md-home"></ons-icon>
      <ons-icon style="color:#0000ff;" icon="md-home"></ons-icon>
    </ons-list-item>

    <ons-list-item>
      <ons-icon size="10px" icon="md-home"></ons-icon>
      <ons-icon size="20px" icon="md-home"></ons-icon>
      <ons-icon size="30px" icon="md-home"></ons-icon>
    </ons-list-item>

    <ons-list-item>
      <ons-icon spin icon="md-spinner"></ons-icon>
      <ons-icon rotate="90" icon="md-home"></ons-icon>
```

```
      <ons-icon rotate="180" icon="md-home"></ons-icon>
      <ons-icon rotate="270" icon="md-home"></ons-icon>
    </ons-list-item>

    <ons-list-item>
      <ons-button id="idHome">
        <ons-icon icon="md-home"></ons-icon> Home
      </ons-button>
    </ons-list-item>

  </ons-list>

  <ons-fab position="bottom right" id="idAction">
    <ons-icon icon="md-plus"></ons-icon>
  </ons-fab>
</ons-page>

<script>
   document.getElementById("idHome").addEventListener("click",
      function(){ons.notification.alert("Home");});
   document.getElementById("idAction").addEventListener("click",
      function(){ons.notification.alert("Action");});
</script>
</body></html>
```

Listing 12.4 onsen_icon.htm File

The icons are displayed here within list entries.

You create an icon by using the ons-icon component. The icon attribute references the name of the icon, and the beginning of this name reveals its origin from widely used icon collections: fa stands for *Font Awesome* and md for *Material Design Iconic Font*.

You use the color CSS property to change the color of the icon, and you use the size attribute to set the size. The spin attribute ensures a permanent rotation of the icon, while you use the rotate attribute to rotate the icon by 90, 180, or 270 degrees, clockwise. You can also integrate icons into other elements, such as here on a button.

The FAB is usually positioned in one of the corners of the screen, usually at the bottom right, although the bottom, right, top, and left values are available for positioning. The FAB covers the screen content at the relevant point. Without special positioning, the FAB is displayed in the normal screen content without covering it. After you tap the FAB, it looks as shown in Figure 12.8.

Figure 12.8 After Tapping FAB

12.2.2   Standard Dialogs

You are already familiar with the simple standard dialog using the ons.notifica-tion.alert() method. There are also three other standard dialogs with the following characteristics:

- The ons.notification.confirm() method corresponds to the confirm() JavaScript method for confirming or rejecting.

- The ons.notification.prompt() method corresponds to the prompt() JavaScript method for entering text.

- The ons.notification.toast() method draws attention to a piece of information. This is pushed in as a message from the bottom of the screen, reminiscent of toast coming out of a toaster.

In Figure 12.9, you can see an application with four buttons that you can use to call the four different standard dialogs.

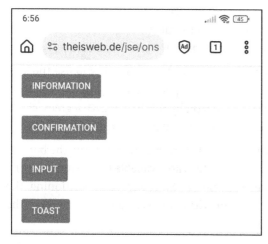

Figure 12.9 Application for Calling Four Standard Dialogs

Now, let's take look at the program for the evaluation of the entries:

```
...
<body>
<ons-page>
  <ons-list>

    <ons-list-item>
      <ons-button id="idInfo">Information</ons-button>
    </ons-list-item>

    <ons-list-item>
      <ons-button id="idConfirmation">Confirmation</ons-button>
    </ons-list-item>

    <ons-list-item>
      <ons-button id="idInput">Input</ons-button>
    </ons-list-item>

    <ons-list-item>
      <ons-button id="idToast">Toast</ons-button>
    </ons-list-item>

  </ons-list>
</ons-page>

<script>
  function info()
  {
    ons.notification.alert("Info");
  }

  const confirmation = function() {
    ons.notification.confirm("Confirmation?")
      .then(function(sch) {
        if(sch==0) ons.notification.alert("Canceled");
        else ons.notification.alert("Confirmed");
      });
  };

  const input = function() {
    ons.notification.prompt("Please make an entry")
      .then(function(tx) {
        if(tx=="") ons.notification.alert("No input");
        else ons.notification.alert("Input: " + tx);
```

```
      });
  };

  function toast()
  {
    ons.notification.toast("The toast", {timeout: 2000});
  }

  document.getElementById("idInfo").addEventListener("click", info);
  document.getElementById("idConfirmation")
      .addEventListener("click", confirmation);
  document.getElementById("idInput").addEventListener("click", input);
  document.getElementById("idToast").addEventListener("click", toast);
</script>
</body></html>
```

Listing 12.5 onsen_dialog.htm File

The buttons for calling the dialogs are displayed here within list items.

The ons.notification.confirm() method calls a message window containing a question (see Figure 12.10). There are two possible buttons that you can use: **OK** and **Cancel**. The internal number of the button that you click gets returned as a value and can be evaluated using the then() method. Here, the value is passed on as a parameter to an anonymous function. After you click the **Cancel** button, this value is 0.

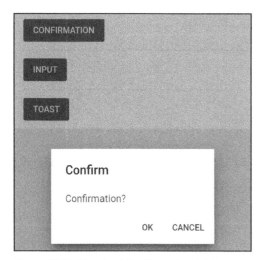

Figure 12.10 Standard Confirmation Dialog

The ons.notification.prompt() method calls a message window containing an input field (see Figure 12.11). A keyboard (not visible here) is also displayed. Once you have made your entry, you must click the **OK** button.

Figure 12.11 Standard Input Dialog

The text of the input is then returned and can also be evaluated using the `then()` method. Here, the text is also passed on as a parameter to an anonymous function. If the text is an empty string, a corresponding message gets displayed; otherwise, the string itself will be displayed.

You can pass the `timeout` parameter with a value in milliseconds to the `ons.notification.toast()` method. In this case, the message will disappear again after the specified time; otherwise, it will remain. In Figure 12.12, you can see the representation of a toast.

Figure 12.12 Standard Toast Dialog

12.2.3   Input Fields

A component of the `ons-input` type corresponds in its function to the `input` HTML element. You can use different types of input fields for numbers, dates and times, phone numbers, URLs, email addresses, files, colors, plaintext, and passwords (see also Chapter 4, Section 4.9).

You can also add designations and placeholders with help texts. Once you select one of the input fields, a corresponding keyboard or selection option will be displayed.

A component of the `ons-search-input` type is more suitable for entering search terms than a component of the `ons-input` type with the `search` subtype.

If you want to select a number from a specific range, a component of the `ons-range` type is again more suitable than a component of the `ons-input` type with the `range` subtype. For more information, see Section 12.2.5.

In Figure 12.13 and in Figure 12.14, you can see examples of the types we've mentioned with placeholders.

There are two buttons at the end of the input form. You can use the first button to display the entered values for checking, as in Figure 12.15. The second button sends the data to a PHP program.

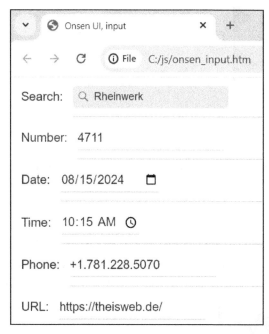

Figure 12.13 Input Fields, Upper Part

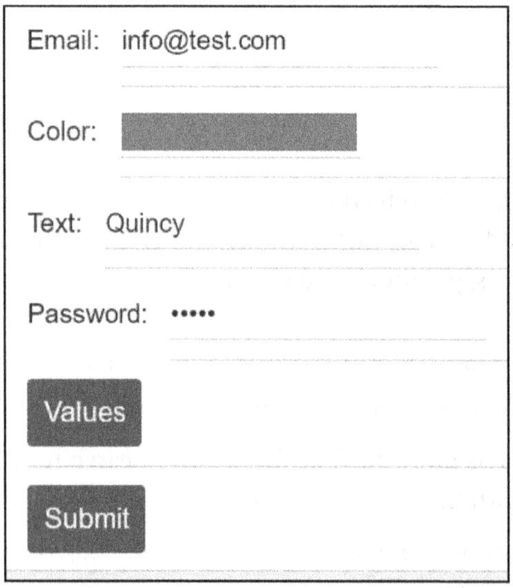

Figure 12.14 Input Fields, Lower Part

```
Alert
Search: Rheinwerk
Number: 4711
Date: 2024-08-15
Time: 10:15
Phone: +1.781.228.5070
URL: https://theisweb.de/
Email: info@test.com
Color: #36aba9
Text: Quincy
Password: bingo

OK
```

Figure 12.15 Entered Values for Checking

Here's the program:

```
... <head> ...
  <script>
    function values()
    {
      ons.notification.alert(
        "Search: " + document.getElementById("se").value
        + "<br>Number: " + document.getElementById("no").value
        + "<br>Date: " + document.getElementById("dt").value
        + "<br>Time: " + document.getElementById("ti").value
        + "<br>Phone: " + document.getElementById("ph").value
        + "<br>URL: " + document.getElementById("ul").value
        + "<br>Email: " + document.getElementById("em").value
        + "<br>Color: " + document.getElementById("co").value
        + "<br>Text: " + document.getElementById("tx").value
        + "<br>Password: " + document.getElementById("pw").value);
    }

    function submit()
    {
      document.getElementById("idForm").submit();
    }
  </script>
</head>
<body>
<form id="idForm" method="post" action="onsen_input.php">
<ons-page>
  <ons-list>

    <ons-list-item>
```

```
    <label for="se" class="left">Search:</label>
    <ons-search-input id="se" name="se" modifier="underbar"
        placeholder="search_term"></ons-search-input>
</ons-list-item>

<ons-list-item>
    <label for="no" class="left">Number:</label>
    <ons-input id="no" modifier="underbar" type="number"
        placeholder="123.456" name="no"></ons-input>
</ons-list-item>

<ons-list-item>
    <label for="dt" class="left">Date:</label>
    <ons-input id="dt" modifier="underbar" type="date"
        name="dt"></ons-input>
</ons-list-item>

<ons-list-item>
    <label for="ti" class="left">Time:</label>
    <ons-input id="ti" modifier="underbar" type="time"
        name="ti"></ons-input>
</ons-list-item>

<ons-list-item>
    <label for="ph" class="left">Phone:</label>
    <ons-input id="ph" modifier="underbar" type="phone"
        placeholder="#49-1234-567890" name="ph"></ons-input>
</ons-list-item>

<ons-list-item>
    <label for="ul" class="left">URL:</label>
    <ons-input id="ul" modifier="underbar" type="url"
        placeholder="www..." name="ul"></ons-input>
</ons-list-item>

<ons-list-item>
    <label for="em" class="left">Email:</label>
    <ons-input id="em" modifier="underbar" type="email"
        placeholder="...@..." name="em"></ons-input>
</ons-list-item>

<ons-list-item>
    <label for="co" class="left">Color:</label>
    <ons-input id="co" name="co" modifier="underbar" type="color"
```

```
        value="#c0c0c0" style="width:150px;"></ons-input>
    </ons-list-item>

    <ons-list-item>
      <label for="tx" class="left">Text:</label>
      <ons-input id="tx" modifier="underbar" placeholder="Enter text"
        name="tx"></ons-input>
    </ons-list-item>

    <ons-list-item>
      <label for="pw" class="left">Password:</label>
      <ons-input id="pw" modifier="underbar" type="password"
        placeholder="(invisible)" name="pw"></ons-input>
    </ons-list-item>

    <ons-list-item>
      <ons-button id="idValues">Values</ons-button>
    </ons-list-item>

    <ons-list-item>
      <ons-button id="idSubmit">Submit</ons-button>
    </ons-list-item>

  </ons-list>
</ons-page>

<script>
  document.getElementById("idValues").addEventListener("click", values);
  document.getElementById("idSubmit").addEventListener("click", submit);
</script>
</form>
</body></html>
```

Listing 12.6 onsen_input.htm File

The entire ons-page component is embedded in a form whose contents are sent to a PHP program. The input fields are displayed here within list items.

You use the id attribute of the input field and the for attribute of the label container to link the input field and the label.

The underbar value for the modifier attribute displays a line below the component for orientation purposes. You can use the value of the placeholder attribute for a placeholder with an explanation of the input field, and the name attribute is required to identify the input field in the responding PHP program.

12.2.4   Selection Fields

A component of the ons-select type corresponds in its function to an HTML selection menu. You can make a default setting using the selected attribute.

Components of the ons-checkbox and ons-radio types correspond in their function to the HTML selection elements, checkbox, and radio button. You can use the checked attribute to set a preselection.

You can use a component of the ons-switch type to create a switch, which provides a choice between two options, such as *on* and *off*. You can operate the switch by dragging or tapping, and here, too, you use the checked attribute for the default setting.

In Figure 12.16 and in Figure 12.17, you can see examples of the selection fields.

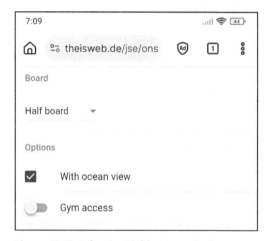

Figure 12.16 Selection Fields, Upper Part

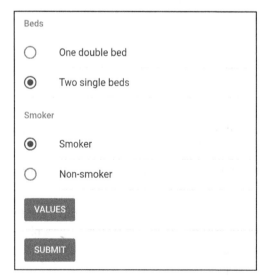

Figure 12.17 Selection Fields, Lower Part

At the end of the input form, there are two buttons for displaying the values (see Figure 12.18) and sending the form.

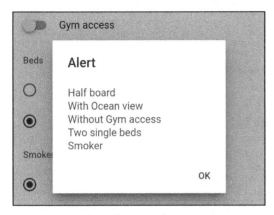

Figure 12.18 Selected Values, for Control Purposes

Here's the program:

```
... <head> ...
  <script>
    function values()
    {
      let output = document.getElementById("bo").value + "<br>";

      output += document.getElementById("ov").checked
        ? "With" : "Without";
      output += " Ocean view<br>";
      output += document.getElementById("gy").checked
        ? "With" : "Without";
      output += " Gym access<br>";

      output += document.getElementById("do").checked
        ? "One double bed<br>" : "Two single beds<br>";
      output += document.getElementById("sm").checked
        ? "Smoker<br>" : "Non-smoker<br>";

      ons.notification.alert(output);
    }

    function submit()
    {
      document.getElementById("idForm").submit();
    }
  </script>
```

```
</head>
<body>
<form id="idForm" method="post" action="onsen_selection.php">
<ons-page>
  <ons-list>

    <ons-list-header>Board</ons-list-header>

    <ons-list-item>
      <ons-select select-id="bo" name="bo">
        <select>
          <option value="Breakfast only"
            selected="selected">Breakfast only</option>
          <option value="Half board">Half board</option>
          <option value="Full board">Full board</option>
        </select>
      </ons-select>
    </ons-list-item>

    <ons-list-header>Options</ons-list-header>

    <ons-list-item>
      <label for="mb">With ocean view</label>
      <label class="left">
        <ons-checkbox id="ov" name="ov"></ons-checkbox>
      </label>
    </ons-list-item>

    <ons-list-item>
      <label for="gy">Gym access</label>
      <label class="left">
        <ons-switch id="gy" name="gy" checked="checked"></ons-switch>
      </label>
    </ons-list-item>

    <ons-list-header>Beds</ons-list-header>

    <ons-list-item>
      <label for="do">One double bed</label>
      <ons-radio class="left" id="do" name="bed"
                 value="One double bed" checked="checked">
      </ons-radio>
    </ons-list-item>
```

```
    <ons-list-item>
      <label for="tw">Two single beds</label>
      <ons-radio class="left" id="tw" name="bed" value="Two single beds">
      </ons-radio>
    </ons-list-item>

    <ons-list-header>Smoker</ons-list-header>

    <ons-list-item>
      <label for="sm">Smoker</label>
      <ons-radio class="left" id="sm" name="smoker" value="Smoker">
      </ons-radio>
    </ons-list-item>

    <ons-list-item>
      <label for="ns">Non-smoker</label>
      <ons-radio class="left" id="ns" name="smoker"
                 value="Non-smoker" checked="checked">
      </ons-radio>
    </ons-list-item>

    <ons-list-item>
      <ons-button id="idValues">Values</ons-button>
    </ons-list-item>

    <ons-list-item>
      <ons-button id="idSubmit">Submit</ons-button>
    </ons-list-item>

  </ons-list>
</ons-page>

<script>
   document.getElementById("idValues").addEventListener("click", values);
   document.getElementById("idSubmit").addEventListener("click", submit);
</script>
</form>
</body></html>
```

Listing 12.7 onsen_selection.htm File

Here, too, the entire ons-page component is embedded in a form whose contents are sent to a PHP program. The selection components are displayed in grouped list items.

A selection menu is identified using the `select-id` attribute, and the `value` property provides the value of the selection menu.

The checkboxes, switches, and radio buttons are identified using the `id` attribute. You use the `?:` ternary operator to evaluate the `checked` property.

The `name` attribute is required for grouping the radio buttons in HTML, while the `name` and `value` attributes are required for the evaluation in PHP.

12.2.5   Selection from a Number Range

A component of the `ons-range` type corresponds in its function to the `input` HTML element with the `range` subtype. The component is used as a slider to select numbers from a range, which can also be numbers with decimal places.

You can define the limits of the number range, the preset value, and the increment for a change. You can operate the slider by dragging or tapping, and in Figure 12.19, you can see an example with three sliders.

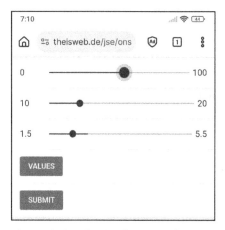

Figure 12.19 Selection from Number Range

A click on the **Values** button displays the current values (see Figure 12.20).

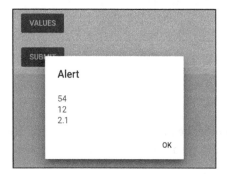

Figure 12.20 Display of Current Values

Here's the program:

```
... <head> ...
  <script>
    function values()
    {
      ons.notification.alert(document.getElementById("rg1").value + "<br>"
        + document.getElementById("rg2").value + "<br>"
        + document.getElementById("rg3").value);
    }

    function submit()
    {
      document.getElementById("idForm").submit();
    }
  </script>
</head>
<body>
<form id="idForm" method="post" action="onsen_range.php">
<ons-page>
  <ons-list>

    <ons-list-item>
        <label for="rg1" class="left">0</label>
        <ons-range id="rg1" name="rg1" value="25" style="width:100%;">
        </ons-range>
        <label for="rg1" class="right">100</label>
    </ons-list-item>

    <ons-list-item>
        <label for="rg2" class="left">10</label>
        <ons-range id="rg2" name="rg2" min="10"
          max="20" step="2" value="12" style="width:100%;">
        </ons-range>
        <label for="rg2" class="right">20</label>
    </ons-list-item>

    <ons-list-item>
        <label for="rg3" class="left">1.5</label>
        <ons-range id="rg3" name="rg3" min="1.5"
          max="5.5" step="0.1" value="2.1" style="width:100%;">
        </ons-range>
        <label for="rg3" class="right">5.5</label>
    </ons-list-item>
```

12

383

```
    <ons-list-item>
      <ons-button id="idValues">Values</ons-button>
    </ons-list-item>

    <ons-list-item>
      <ons-button id="idSubmit">Submit</ons-button>
    </ons-list-item>

  </ons-list>
</ons-page>

<script>
  document.getElementById("idValues").addEventListener("click", values);
  document.getElementById("idSubmit").addEventListener("click", submit);
</script>
</form>
</body></html>
```

Listing 12.8 onsen_range.htm File

The entire ons-page component is also embedded in a form whose content is sent to a PHP program. The sliders and their range limits are displayed in list items.

By default, the range extends from 0 to 100, with an increment of 1. You can use the min and max attributes to change the range limits and the step attribute to change the increment. You can also use numbers with decimal places, and you can use the value attribute for the default setting.

A slider is identified using the value of the id attribute. The value property provides the current value.

> **Note**
>
> Bonus chapter 5 in the downloadable materials for this book (found at *www.rhein-werk-computing.com/5875*) describes three other programs that use the Onsen UI library. They describe different methods of displaying documents with multiple pages.

Chapter 13
Mathematical Expressions Using MML and MathJax

You can create mathematical formulas dynamically and interactively using MathML, MathJax, and JavaScript.

You can use *Mathematical Markup Language* (MathML) to integrate mathematical expressions into documents. In collaboration with JavaScript, you have further possibilities for dynamic and interactive design. The W3C published its recommendation for the second edition of the MathML 3.0 standard in April 2014. MathML is part of HTML version 5 and has been an ISO standard since 2015.

Unfortunately, MathML doesn't immediately get processed in certain browsers. However, the JavaScript *MathJax* library (*https://www.mathjax.org*) enables the implementation of MathML in all modern browsers, so we include the MathJax library in the sample programs in this chapter.

13.1 Basic Elements

The following program shows important basic elements of the structure of mathematical expressions:

```
... <head> ...
   <script id="MathJax-script" async src="https://cdn.jsdelivr.net/
         npm/mathjax@3/es5/tex-mml-chtml.js"></script>
</head>
<body>
<p> Identifier, Operator, Number:
   <math>
       <mi>a</mi>
       <mo>+</mo>
       <mn>2.5</mn>
       <mo>=</mo>
       <mn>3</mn>
       <mi>a</mi>
       <mi>b</mi>
   </math>
```

```
</p>
<p> Square root:
    <math>
        <msqrt> <mi>a</mi> <mo>+</mo> <mn>2.5</mn> </msqrt>
    </math> </p>
<p> Root:
    <math> <mroot> <mi>b</mi> <mn>3</mn> </mroot> </math> </p>
<p> Index:
    <math> <msub> <mi>a</mi> <mn>2</mn> </msub> </math> </p>
<p> Exponent:
    <math> <msup> <mi>b</mi> <mn>-3.5</mn> </msup> </math> </p>
<p> Index and Exponent:
    <math>
        <msubsup> <mi>c</mi> <mn>0</mn> <mn>2</mn> </msubsup>
    </math> </p>
<p> Format:
    <math mathbackground="#909090" mathcolor="#ffffff" mathsize="16pt">
        <msqrt> <mi>a</mi> <mo>+</mo> <mn>2.5</mn> </msqrt>
    </math> </p>
</body></html>
```

Listing 13.1 mm_basic_element.htm File

The online version of the MathJax library is included in the header of the file. The address should be uninterrupted and within one line. There are only two lines here for printing reasons. We use the `async` keyword to ensure that we make the request for the online library in parallel with the construction of the rest of the document. The integration looks the same for the other programs in this chapter, and we therefore no longer show it there.

All elements of mathematical expressions and formulas are located within the `<math>` container, which in turn is located within the normal text and HTML code of a page.

You write the elements one after the other in a simple mathematical expression. Mathematical *identifiers* are stored in the `<mi>` container, operators in the `<mo>` container, and numerical values in the `<mn>` container. In mathematics, the 3ab expression represents the multiplication 3 * a * b, by default without any multiplication sign. The a and b identifiers are separate, and each is in its own `<mi>` container.

You write many of the following MathML containers in a more compact form for a better overview, so the embedded containers are no longer in their own lines in the file.

You use the `<msqrt>` container to display a square root. You write the so-called *radicand* under a root symbol; it is the value from which the root is taken. In the case of the square root, this can be a single identifier or a number, but it can also be a longer expression whose elements are written one after the other.

You use the `<mroot>` container to display a general root. In contrast to the `<msqrt>` container, we must write two elements within the `<mroot>` container: first, the radicand (here, b), and then, the *root exponent* (here, 3). If one of the two elements consists of a longer expression, this expression must first be summarized. You can do that using the `<mrow>` container (Section 13.3), and this rule for summarizing also applies to many containers in the examples that follow.

You write the index of an identifier by using the `<msub>` container. There must be two elements within the container: first, the identifier, and then, the index. You create an expression with an exponent by using the `<msup>` container, and there must also be two elements within the container: the base and the exponent. You write an expression that includes both an index and an exponent by using the `<msubsup>` container. There must be three elements within the container: the identifier, the index, and the exponent.

For the `<math>` container and many containers for the individual elements, you can set the background color, font color, and font size by using the `mathbackground`, `mathcolor`, and `mathsize` attributes.

The mathematical expressions are shown in Figure 13.1.

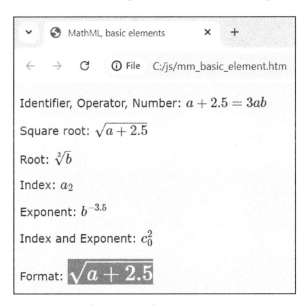

Figure 13.1 MathML, Basic Elements

13.2 Parentheses and Tables

Within mathematical expressions, subexpressions are grouped together using parentheses so that they are calculated with priority. MathML uses parentheses to visually represent this order of operations. Parentheses are also necessary to summarize expressions, and they are required to represent a vector or a matrix. You can use any parenthesis characters and work with multiple parenthesis levels.

You can use tables to provide clear representation of two-dimensional structures. In this example, a matrix and a system of equations are displayed using tables.

Let's first take a look at the sample program:

```
...
<body>
<p> Single parentheses:
    <math> <mfenced separators="">
            <mi>a</mi> <mo>+</mo> <mi>b</mi>
        </mfenced> </math> </p>
<p> Double parentheses:
    <math>
      <mfenced open="{" close="}" separators="">
        <mfenced separators=""> <mi>a</mi> <mo>+</mo>
                                <mi>b</mi> </mfenced>
        <mfenced separators=""> <mi>c</mi> <mo>+</mo>
                                <mi>d</mi> </mfenced>
      </mfenced>
    </math> </p>
<p> Table, for equations:
    <math> <mtable>
            <mtr>
                <mtd> <mn>1.5</mn> <mo>+</mo> <mi>a</mi> </mtd>
                <mtd> <mo>=</mo> </mtd>
                <mtd> <mi>b</mi> </mtd>
            </mtr>
            <mtr>
                <mtd> <mi>a</mi> </mtd>
                <mtd> <mo>=</mo> </mtd>
                <mtd> <mn>4.2</mn> <mo>-</mo> <mi>b</mi> </mtd>
            </mtr>
        </mtable> </math> </p>
<p> Table, for matrix:
    <math> <mi>X</mi> <mo>=</mo>
      <mfenced separators="">
        <mtable>
          <mtr> <mtd> <mn>1.5</mn> </mtd> <mtd> <mi>a</mi>   </mtd> </mtr>
          <mtr> <mtd> <mi>b</mi>   </mtd> <mtd> <mn>2.8</mn> </mtd> </mtr>
        </mtable>
      </mfenced>
```

```
    </math> </p>
</body></html>
```

Listing 13.2 mm_parenthesis_table.htm File

You can use the `<mfenced>` container to display a *fenced* pair of parentheses. The elements are written one after the other within the container, and the value of the separators attribute stands for the string that is used to separate the individual elements. The string is a comma character by default. In this example, it's an empty string.

You can use nested `<mfenced>` containers to implement multiple parenthesis levels, and the `open` and `close` attributes allow you to define two strings for each pair of parentheses that are to be used as opening or closing parentheses.

The structure and purpose of MathML tables are similar to the structure and purpose of HTML tables. There are the `<mtable>` containers for the table, `<mtr>` for the rows within a table, and `<mtd>` for the data cells within a row.

In the first table example, the left-hand sides, the equal signs, and the right-hand sides of the two equations are displayed one above the other thanks to the table. You can map an entire table within parentheses, as in the second table example. This allows you to display the data as a matrix.

In Figure 13.2, you can see the mathematical expressions with parentheses and tables.

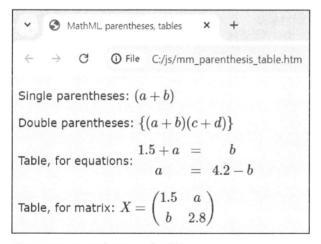

Figure 13.2 Parentheses and Tables

13.3 Summarizing Expressions

The `<mrow>` container allows you to summarize a longer expression into a single element. Many containers require this procedure for their elements.

You can use the <munderover> container to enter the necessary information below and above characters. These can be the symbols for an integral, a sum, or a product, for example.

Here, you can see some examples:

```
...
<body>
<p> Create summary, for square root:
    <math> <mroot>
            <mrow> <mi>a</mi> <mo>+</mo> <mn>2.5</mn> </mrow> <mn>3</mn>
        </mroot> </math> </p>
<p> Integral:
    <math> <munderover>
            <mo>&int;</mo> <mn>0</mn> <mn>1</mn>
        </munderover>
        <mi>x</mi> <mi>d</mi> <mi>x</mi> </math> </p>
<p> Create summary, for sum:
    <math> <munderover>
            <mo>&sum;</mo>
            <mrow> <mi>k</mi> <mo>=</mo> <mn>3</mn> </mrow>
            <mn>5</mn>
        </munderover>
        <mi>k</mi> <mo>=</mo>
            <mn>3</mn> <mo>+</mo> <mn>4</mn> <mo>+</mo> <mn>5</mn>
        <mo>=</mo> <mn>12</mn> </math> </p>
<p> Create summary, for product:
    <math> <munderover>
            <mo>&prod;</mo>
            <mrow> <mi>k</mi> <mo>=</mo> <mn>3</mn> </mrow>
            <mn>5</mn>
        </munderover>
        <mi>k</mi> <mo>=</mo> <mn>3</mn> <mo>&sdot;</mo>
            <mn>4</mn> <mo>&sdot;</mo> <mn>5</mn>
        <mo>=</mo> <mn>60</mn> </math> </p>
</body></html>
```

Listing 13.3 mm_summary.htm File

The radicand under the root is a longer expression, and you therefore write it within an <mrow> container. Alternatively, you could use an <mfenced> container without visible parentheses.

The HTML special character ∫ creates an integral symbol. There must be three elements in a <munderover> container: the actual main element, the information below the element, and the information above the element. If you were to place the dx string in a common <mi> container in the example, there would be no mathematical representation, so it's better to place the d and x characters in their own <mi> containers.

The HTML special characters ∑ and ∏ create a summation symbol and a product symbol, respectively. The lower specification in the <munderover> container consists of a longer expression and must therefore be summarized. The multiplication sign within the intermediate step for calculating the product is displayed using the special HTML character ⋅ for clarification.

In Figure 13.3, you can see the summarized expressions, the integral, the sum, and the product.

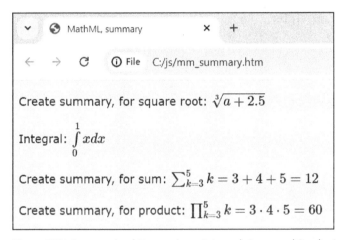

Figure 13.3 Summarized Expressions, Integral, Sum, and Product

13.4   Fractions

You can display fractions by using the <mfrac> container. Usually, you would use the horizontal fraction bar, but using the slanted fraction bar is also possible. You can also nest expressions with fractions, and in this way, you have the option of representing a fraction in the numerator or denominator. Here, you can see some examples:

```
...
<body>
<p> Fraction, horizontal:
    <math>
        <mfrac>
```

```
            <mi>a</mi> <mn>3</mn>
        </mfrac>
    </math> </p>
<p> Fraction, slanted:
    <math>
        <mfrac bevelled="true"> <mi>a</mi> <mn>3</mn> </mfrac>
    </math> </p>
<p> Fraction, horizontal, with expression:
    <math>
        <mfrac>
          <mrow> <mi>a</mi> <mo>+</mo> <mi>b</mi> </mrow>
          <mn>2</mn>
        </mfrac>
    </math> </p>
<p> Fraction, horizontal, multiple:
    <math mathsize="16pt">
        <mfrac>
          <mrow>
            <mfrac>
              <mrow> <mi>a</mi> <mo>+</mo> <mi>b</mi> </mrow>
              <mn>2</mn>
            </mfrac>
            <mo>+</mo>
            <mn>3</mn>
          </mrow>
          <mi>c</mi>
        </mfrac>
    </math> </p>
</body></html>
```

Listing 13.4 mm_fraction.htm File

As expected, there must be two elements in an <mfrac> container: the numerator and the denominator. The true value for the bevelled attribute creates a slanted fraction bar instead of a horizontal fraction bar. You can convert a longer expression in the numerator or denominator using a <mrow> container.

In the last example, there's a longer expression in the numerator, which in turn consists of a fraction and other elements. A higher value for the font size is recommended here.

You can see the various fractions in Figure 13.4.

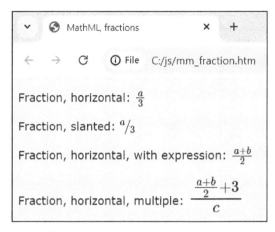

Figure 13.4 Fractions

13.5   Mathematical Symbols

There's a whole range of special HTML characters that are suitable for displaying an operator within a MathML expression, and you can see some of them together with a brief explanation in the following two-part program:

```
...
<body>
<p>Comparison:<br>
    <math> <mo>&sim;</mo> </math> Similar to<br>
    <math> <mo>&asymp;</mo> </math> Almost equal, rounded<br>
    <math> <mo>&ne;</mo> </math> Unequal to<br>
    <math> <mo>&lt;</mo> </math> Less than<br>
    <math> <mo>&gt;</mo> </math> Greater than<br>
    <math> <mo>&le;</mo> </math> Less than or equal to<br>
    <math> <mo>&ge;</mo> </math> Greater than or equal to</p>

<p>Logic:<br>
    <math> <mo>&and;</mo> </math> Logical And<br>
    <math> <mo>&or;</mo> </math> Logical Or<br>
    <math> <mo>&not;</mo> </math> Logical Not<br>
    <math> <mo>&rarr;</mo> </math> If, then (conditional)<br>
    <math> <mo>&harr;</mo> </math> Exactly when (biconditional)</p>
...
```

Listing 13.5 mm_symbols.htm File, Part 1 of 2

You can see the display in Figure 13.5.

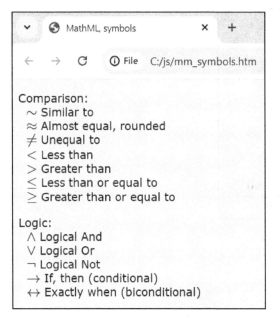

Figure 13.5 Mathematical Symbols, Part 1

Here's the second part of the program:

```
...
<p>Sets:<br>
    <math> <mo>&isin;</mo> </math> Element of<br>
    <math> <mo>&notin;</mo> </math> No element of<br>
    <math> <mo>&cap;</mo> </math> Intersection<br>
    <math> <mo>&cup;</mo> </math> Union</p>

<p>Angles:<br>
    <math> <mo>&deg;</mo> </math> Degree<br>
    <math> <mo>&prime;</mo> </math> Degree minute<br>
    <math> <mo>&Prime;</mo> </math> Degree second<br>
    <math> <mo>&alpha;</mo> </math> Alpha<br>
    <math> <mo>&beta;</mo> </math> Beta<br>
    <math> <mo>&gamma;</mo> </math> Gamma</p>

<p>Other symbols:<br>
    <math> <mo>&Delta;</mo> </math> Delta, difference<br>
    <math> <mo>&pi;</mo> </math> Pi<br>
    <math> <mo>&micro;</mo> </math> Micro<br>
    <math> <mo>&plusmn;</mo> </math> Plus-minus<br>
```

```
    <math> <mo>&infin;</mo> </math> Infinity</p>
</body></html>
```

Listing 13.6 mm_symbols.htm File, Part 2 of 2

The characters are displayed in Figure 13.6.

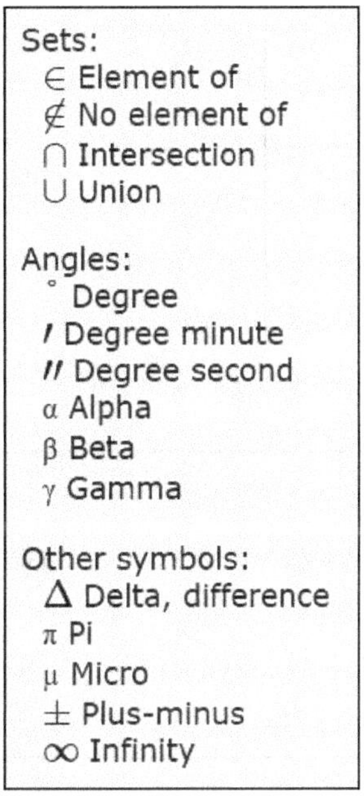

```
Sets:
  ∈ Element of
  ∉ No element of
  ∩ Intersection
  ∪ Union

Angles:
  ° Degree
  ′ Degree minute
  ″ Degree second
  α Alpha
  β Beta
  γ Gamma

Other symbols:
  Δ Delta, difference
  π Pi
  μ Micro
  ± Plus-minus
  ∞ Infinity
```

Figure 13.6 Mathematical Symbols, Part 2

13.6   Dynamically Generated Expressions

The dynamic insertion of mathematical expressions into documents that have already been fully loaded is very time-consuming. For this reason, I use two separate documents in this section: the location object and the transfer of a search string (see also Chapter 4, Section 4.8).

Data is entered in the first document and is then sent to the second document using JavaScript, where it will be fully available when the document gets loaded. The data can then be inserted into the mathematical expressions, again using JavaScript.

In this example, the scalar product of two vectors with three components each is calculated. The scalar product corresponds to the sum of the products of the related components, and in this example, the result is as follows:

```
sp = 3.2 * (-2.6) + (-2.7) * 0.6 + 1.9 * 3.4 = -3.48
```

You enter components and separate them with spaces (see Figure 13.7). After you send them, the result is calculated and the entire expression gets output (see Figure 13.8).

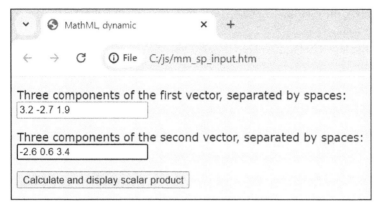

Figure 13.7 Entering Components

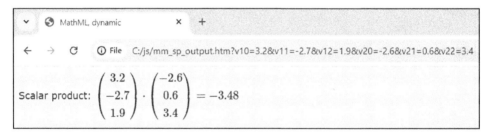

Figure 13.8 Calculation of Result and Representation

You can apply the technique shown here to more complex examples. These could include a scalar product of two vectors with any number of components or the multiplication of two matrixes.

First, here's the program for input:

```
... <head> ...
  <script>
    function calculate()
    {
        const tx1 = document.getElementById("idVector1").value;
        const v1 = tx1.split(" ");
        const tx2 = document.getElementById("idVector2").value;
        const v2 = tx2.split(" ");
        if(v1.length != 3 || v2.length != 3)
        {
            alert("There must be three components in each case");
            return;
        }
    }
```

```
            location.href = "mm_sp_output.htm"
               + "?v10=" + v1[0] + "&v11=" + v1[1] + "&v12=" + v1[2]
               + "&v20=" + v2[0] + "&v21=" + v2[1] + "&v22=" + v2[2];
      }
   </script>
</head>
<body>
<p>Three components of the first vector, separated by
   spaces:<br><input size="20" id="idVector1"></p>
<p>Three components of the second vector, separated by
   spaces:<br><input size="20" id="idVector2"></p>
<p><input type="button" id="idCalculate"
   value="Calculate and display scalar product"></p>
<script>
   document.getElementById("idCalculate").addEventListener("click", calculate);
</script>
</body></html>
```

Listing 13.7 mm_sp_input.htm File

After you enter the data and click the button, both input strings are separated by a space and saved in a field. If one of the fields doesn't contain exactly three elements, the program will be exited. Using the required special characters ?, &, and = as well as the contents of the two fields, the search string is put together and the second document gets called.

Now, here's the program for the evaluation:

```
...
<body>
<script>
   const v = new Array();
   const search_field = location.search.split("&");
   for(let i=0; i<search_field.length; i++)
   {
      if(i==0)
         search_field[0] = search_field[0].substr(1);
      const parts = search_field[i].split("=");
      v.push(parts[1]);
   }

   document.write("Scalar product: <math>");
```

```
    document.write("<mfenced separators=''> <mtable>");
    for(let i=0; i<3; i++)
        document.write("<mtr> <mtd> <mn>" + v[i] + "</mn> </mtd> </mtr>");
    document.write("</mtable> </mfenced>");

    document.write("<mo>&sdot;</mo>");

    document.write("<mfenced separators=''> <mtable>");
    for(let i=0; i<3; i++)
        document.write("<mtr> <mtd> <mn>" + v[i+3] + "</mn> </mtd> </mtr>");
    document.write("</mtable> </mfenced>");

    let sp = 0;
    for(let i=0; i<3; i++)
        sp += v[i] * v[i+3];
    document.write("<mo>=</mo> <mn>" + sp.toFixed(2) + "</mn>");

    document.write("</math>");
</script>
</body></html>
```

Listing 13.8 mm_sp_output File

The search string is broken down again using the special characters, and the total of six components of the two vectors are stored in one field. The <mfenced> and <mtable> containers are used to display the values of the two vectors, and the scalar product is calculated and also displayed.

Chapter 14
Sample Projects

Larger projects demonstrate the interaction of the various capabilities of JavaScript particularly well.

In this chapter, I want to describe a number of larger sample projects. First, you'll see three practice-oriented projects:

- Calculation of the value of a financial investment
- Calculation and assessment of values for the fitness area
- Registration for a fun run, with a check of the input values

I follow these with three projects from the games sector:

- A *solitaire* game in which playing cards have to be placed in a certain order after calm deliberation
- The *Concentration* game in a version for two players
- The dexterity game *Snake*, in which players control an object on the screen

These programs also require careful planning during development, and the end products are just as popular as the serious programs.

You'll find all programs as a bonus in the downloadable materials for this book, with many explanatory comments. You can generally use the programs to understand the capabilities of JavaScript in interaction, and you can also expand the programs with your own ideas.

14.1 Financial Investment

This program is available in the *investment.htm* file, and the various conditions offered by the bank are displayed at the bottom of the page. You can enter the investment amount and the desired term, and once you click the **Calculate** button, the program processes the input values and returns the amount at the end of the runtime (see Figure 14.1).

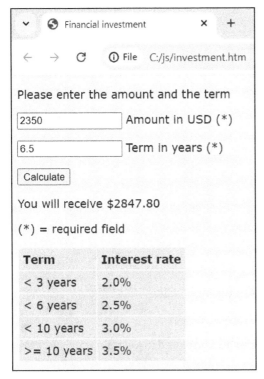

Figure 14.1 Financial Investment

14.2   Fitness Values

You can find this program in the *fitness.htm* file, and you can enter the personal values—age, height, weight, resting heart rate, and gender—at the top of the page.

When you click the **Calculate** button, the program checks the input values and uses them to calculate the body mass index and the training heart rates according to the Karvonen formula (see Figure 14.2).

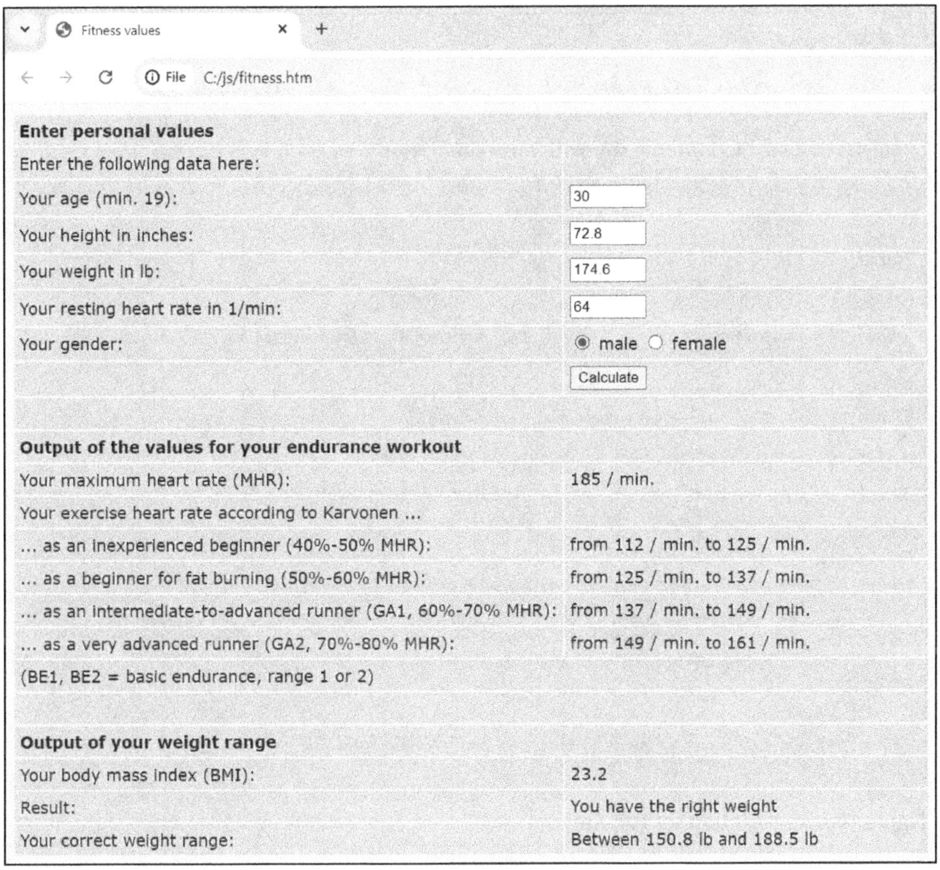

Fitness values

Enter personal values

Enter the following data here:

Your age (min. 19):	30
Your height in inches:	72.8
Your weight in lb:	174.6
Your resting heart rate in 1/min:	64
Your gender:	◉ male ○ female
	Calculate

Output of the values for your endurance workout

Your maximum heart rate (MHR):	185 / min.
Your exercise heart rate according to Karvonen ...	
... as an inexperienced beginner (40%-50% MHR):	from 112 / min. to 125 / min.
... as a beginner for fat burning (50%-60% MHR):	from 125 / min. to 137 / min.
... as an intermediate-to-advanced runner (GA1, 60%-70% MHR):	from 137 / min. to 149 / min.
... as a very advanced runner (GA2, 70%-80% MHR):	from 149 / min. to 161 / min.
(BE1, BE2 = basic endurance, range 1 or 2)	

Output of your weight range

Your body mass index (BMI):	23.2
Result:	You have the right weight
Your correct weight range:	Between 150.8 lb and 188.5 lb

Figure 14.2 Fitness Values

14.3   Fun Run

The program is available in the *fun_run.htm* file, and you can enter the personal data (see Figure 14.3). The last name, first name, and consent to data storage are mandatory fields, and you must choose the gender, year of birth, and distance so that they match each other. In Figure 14.4, you can see the expanded selection menu for the various distances that can be run.

In a registration for a real fun run, the data is then stored in a MySQL database using PHP, and you can view a list of registered participants for checking purposes. Organizers have the option of exporting all data from the database to a CSV file in a protected area (e.g., for Microsoft Excel). This simplifies the evaluation of the runs and the printing of the results lists and certificates.

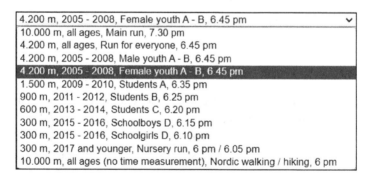

Figure 14.3 Registration for Fun Run

4.200 m, 2005 - 2008, Female youth A - B, 6.45 pm	⌄
10.000 m, all ages, Main run, 7.30 pm	
4.200 m, all ages, Run for everyone, 6.45 pm	
4.200 m, 2005 - 2008, Male youth A - B, 6.45 pm	
4.200 m, 2005 - 2008, Female youth A - B, 6.45 pm	
1.500 m, 2009 - 2010, Students A, 6.35 pm	
900 m, 2011 - 2012, Students B, 6.25 pm	
600 m, 2013 - 2014, Students C, 6.20 pm	
300 m, 2015 - 2016, Schoolboys D, 6.15 pm	
300 m, 2015 - 2016, Schoolgirls D, 6.10 pm	
300 m, 2017 and younger, Nursery run, 6 pm / 6.05 pm	
10.000 m, all ages (no time measurement), Nordic walking / hiking, 6 pm	

Figure 14.4 Various Distances

14.4   Solitaire

This program is located in the *solitaire.htm* file in the *solitaire* subdirectory. You can also find the corresponding image files there, including those for the 52 cards.

In Figure 14.5, you can see the cards in the original order, which is generated randomly at the beginning of the game. After clicking the **Rules** button, you'll see a description of the process (see Figure 14.6).

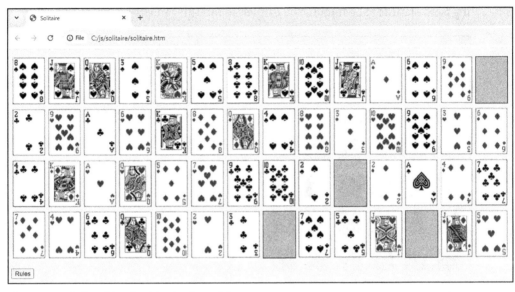

Figure 14.5 Order of Cards at Beginning of Game

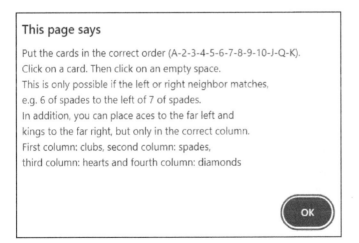

Figure 14.6 Rules of Game

At the end, all the cards should be in the correct order.

14.5   Concentration

You can find this program in the *concentration.htm* file in the *concentration* subdirectory, and you can also find the associated image files there (including for the 18 matching pairs of cards to be found). You can see an interim game result in Figure 14.7.

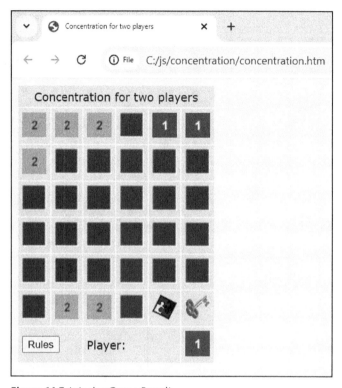

Figure 14.7 Interim Game Result

Player 2 has already found three pairs of cards, while player 1 has only found one pair. It's player 1's turn now, and he has just flipped the second card. Both players now have 2 to 3 seconds to memorize the position of the two face-up cards, and you can see the backs of all remaining cards. After you click the **Rules** button, a description of the process gets displayed (see Figure 14.8).

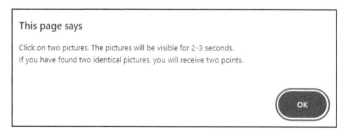

Figure 14.8 Game Rules

Players 1 and 2 normally alternate taking turns, but when a player finds a matching pair of cards, that player can immediately take another turn.

14.6   Snake

You can find this program in the *snake.htm* file. Immediately after you call the program, the snake moves automatically across the screen, and you can control its direction of movement by using the four arrow keys so that the snake can catch its prey (see Figure 14.9). The game ends as soon as the snake hits one of the surrounding walls.

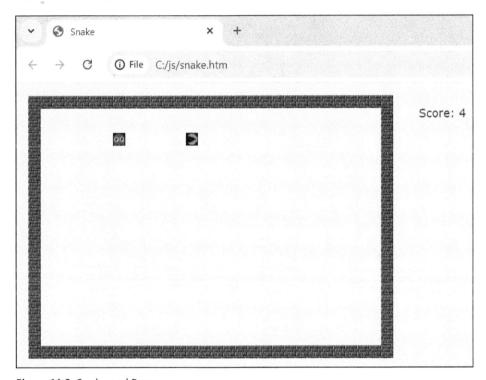

Figure 14.9 Snake and Prey

As soon as the snake reaches its prey, one point gets added to the score. Then the prey appears at a new, random position, and the snake tries to reach it again. At the same time, the snake becomes faster and therefore more difficult to control, so that the game runs towards its natural end as the number of points increases.

Chapter 15
Media, Drawings, and Sensors

Since its version 5 came out, HTML has provided additional capabilities,
also in combination with JavaScript.

Since version 5 came out, HTML has provided some additional capabilities that are also interesting in connection with JavaScript. These capabilities include the playback of audio and video files, canvas drawings, and the use of receivers or sensors for location data, position, and acceleration of a mobile device.

15.1 Playing Media Files

The audio and video tags make it much easier to play audio and video files, compared to the options that existed prior to version 5 of HTML.

15.1.1 Audio Files

In your document, you can use the audio tag to deploy an audio player that has some display and control elements: a play/pause button, an analog and digital progress indicator, and a volume control. If required, you can also hide an audio player or control it by using events.

The following is an example with two Waveform Audio File Format (WAV) files and a total of three audio players (see Figure 15.1).

You can operate the first audio player in the usual way, but the second audio player is not visible, and you can only control it by using events and JavaScript.

There are some more options shown for the third player. You can do the following:

- Pause or continue the playback process.
- Continue the playback process from a certain point.
- Switch the endless repetition on and off.
- Adjust the volume.
- Switch between two WAV files.

Figure 15.1 Three Audio Players

Let's first look at the lower part of the document:

```
...
<body>
    <p>Audio player 1:</p>
    <p><audio src="guitar.wav" controls></audio></p>
    <p> </p>

    <!-- Audioplayer 2: -->
    <audio id="idAudioHidden" src="chord.wav" hidden></audio>
    <p><input id="idHidden" type="button"
        value="Play hidden audio player"></p>
    <p> </p>

    <p>Audio player 3:</p>
    <p>
        <audio id="idAudioThree" src="guitar.wav" loop></audio>
        <input id="idPlay" type="button" value="Play">
        <input id="idPause" type="button" value="Pause">
        <input id="idSet" type="button" value="Set to 1.5">
        <input id="idLoop" type="checkbox" checked="checked"> Loop</p>
    <p><input id="idVolumeUp" type="button" value="Volume up">
```

```
    <input id="idVolumeDown" type="button" value="Volume down">
    <span id="idDisplayThree">(100%)</span></p>
  <p><input id="idGuitar" name="selection" type="radio"
          value="Guitar" checked="checked"> guitar.wav
  <input id="idChord" name="selection" type="radio"
        value="Chord"> chord.wav</p>

<script>
    const audioHidden =
      document.getElementById("idAudioHidden");
    document.getElementById("idHidden").addEventListener
      ("click", function(){audioHidden.play();});

    const audioThree = document.getElementById("idAudioThree");
    document.getElementById("idPlay").addEventListener
      ("click", function(){audioThree.play();});
    document.getElementById("idPause").addEventListener
      ("click", function(){audioThree.pause();});
    document.getElementById("idSet").addEventListener
      ("click", function(){audioThree.currentTime=1.5;});
    document.getElementById("idLoop").addEventListener
      ("click", function(){audioThree.loop = !audioThree.loop;});
    document.getElementById("idVolumeUp").addEventListener
      ("click", function(){volume(0.2);});
    document.getElementById("idVolumeDown").addEventListener
      ("click", function(){volume(-0.2);});
    document.getElementById("idGuitar").addEventListener
      ("click", function(){change("guitar");});
    document.getElementById("idChord").addEventListener
      ("click", function(){change("chord");});
  </script>
</body></html>
```

Listing 15.1 audio.htm File, Audio Elements, and Event Handlers

Some notes on the three audio players:

- For the first player, you use the value of the src attribute to specify the name of the audio file, and you use the controls attribute to display the operating elements.
- The second player has the hidden attribute in addition to the src attribute. This ensures that although it's present and usable, it's not visible. The unique ID establishes the connection to the controller using JavaScript.
- The third player has the loop attribute entered, and this starts the endless playback process.

This is followed by a button for operating the hidden player, and a link to this player is set in the JavaScript section. Clicking the **Play hidden audio player** button calls the play() method for this player.

There are a total of five buttons, a checkbox, and a group of two radio buttons that you use to operate the third player. A reference to this player is also set in the JavaScript section. Clicking the **Play** and **Pause** buttons calls the play() and pause() methods. You use the currentTime property to determine or set the current position of the playback process, and you use the checkbox to activate or deactivate the Boolean property loop for endless repetition.

You can't preset the volume value. Initially, you'll hear the sound at the volume that is currently set on the output device (PC, smartphone, etc.), which corresponds to the value 1 for the volume property. You call the volume() function by using the **Volume up** and **Volume down** buttons. A value that changes the volume is passed as a parameter, and the changed volume can only be between 0 and 1. You use the two radio buttons to select the audio file you want to play.

Now, here's the upper part of the document:

```
...  <head> ...
   <script>
      function volume(value)
      {
         audioThree.volume += value;
         document.getElementById("idDisplayThree")
            .firstChild.nodeValue =
            "(" + (100 * audioThree.volume).toFixed(0) + "%)";
      }

      function change(filename)
      {
         audioThree.src = filename + ".wav";
         audioThree.load();
         audioThree.play();
      }
   </script>
</head>
...
```

Listing 15.2 audio.htm File

You can change the value of the volume property in the volume() function, and that will change the percentage of the volume that you could hear at the beginning from 100% to a new percentage.

In the change() function, you give the src property a new value by using the function parameter. This property contains the name of the audio file for the player, and the load() method loads the corresponding audio file, which you can play via play().

15.1.2 Video Files

You can play video files using the video tag in your browser. Not every browser recognizes all video formats, so it makes sense to convert a video file into multiple formats and integrate all formats using the source tag. The first format that is recognized will then be used, and there are a number of freeware programs that allow you to convert from one format to another.

In addition to the attributes you learned about when tagging audio content, there are the width and height attributes, which you can use to reserve and define the area for the video within the website during the loading process. In JavaScript, you can use the same properties as for audio files.

In Figure 15.2, you can see the display of a short video. If the video is not currently running or the mouse cursor is hovering over the video, a control bar gets displayed.

Figure 15.2 Video File with Control Bar

Here's the program code:

```
...
<body>
    <video width="400" height="300" controls loop>
```

```
    <source src="shadow1.mp4"
            type='video/mp4; codecs="avc1.42E01E, mp4a.40.2"'>
    <source src="shadow1.ogv"
            type='video/ogg; codecs="theora, vorbis"'>
    <source src="shadow1.webm"
            type='video/webm; codecs="vp8, vorbis"'>
  </video>
</body></html>
```

Listing 15.3 video.htm File

An area with a size of 400 × 300 pixels is reserved and defined for the video. There can be multiple source tags within a video container, and in each case, you want to use the src attribute to specify the name of the file being played.

Specifying the MIME type, including the codec, makes it easier for the browser to assign the file. The codec identifies the process by which the analog video recording is digitized.

15.2   Canvas

The canvas tag creates a blank *canvas* within a website. You can then use JavaScript to create drawings, insert images, or output formatted text, for example.

15.2.1   Drawings

In Figure 15.3, you can see some drawing elements that are generated each time you click one of the buttons. You use the **Clear** button to clear the canvas.

Let's look at the lower part of the document first:

```
...
<body>
  <canvas id="idCanvas" width="400" height="300"
          style="border:solid 1px #000000;"></canvas>
  <p><input id="idRect" type="button" value="Rectangle">
    <input id="idArc" type="button" value="Arc">
    <input id="idLine" type="button" value="Line">
    <input id="idClear" type="button" value="Clear"></p>
  <p><input id="idGradientLin" type="button"
            value="Linear gradient">
    <input id="idGradientRad" type="button"
            value="Radial gradient"></p>
  <script>
    const c = document.getElementById("idCanvas");
```

```
        const ct = c.getContext("2d");
        ct.fillStyle = "#f0f0f0";
        ct.lineWidth = 1;
        ct.strokeStyle = "#000000";

        document.getElementById("idRect").addEventListener("click", rectangle);
        document.getElementById("idArc").addEventListener("click", arc);
        document.getElementById("idLine").addEventListener("click", line);
        document.getElementById("idClear").addEventListener("click", clear);
        document.getElementById("idGradientLin").addEventListener("click", gLinear);
        document.getElementById("idGradientRad").addEventListener("click", gRadial);
    </script>
</body></html>
```

Listing 15.4 canvas.htm File, Lower Part

The canvas has a size of 400 × 300 pixels, and it's surrounded by a thin black frame for better visibility within the page. Then, the six buttons follow.

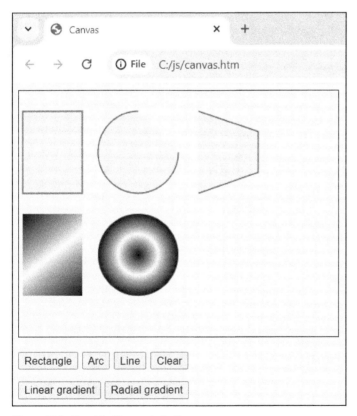

Figure 15.3 Graphic Elements in Canvas

In the JavaScript section, the c variable is introduced as a reference to the canvas element. The getContext() method returns an object that can be used for drawing purposes. Some default settings are made and apply until other settings are made; after that, the new settings apply:

- The fillStyle property identifies the fill type and color of an object. By default, objects are filled evenly with black color. The color gray is set here, but you can specify a pattern or a gradient (i.e., a color gradient).

- The lineWidth property stands for the line thickness. Its default value is 1.

- The strokeStyle property identifies the line style and color. By default, lines are drawn continuously with a black color, but you can set a different line type.

Finally, there are the event handlers for the six buttons.

Let's now look at the upper part of the document with the functions for creating a rectangle, an arc, and a line:

```
... <head> ...
  <script>
    function rectangle()
    {
      ct.fillRect(5, 25, 75, 100);
      ct.strokeRect(5, 25, 75, 100);
    }

    function pathArc()
    {
      ct.beginPath();
      ct.arc(150, 75, 50, 0, 1.5 * Math.PI);
      ct.fill();
      ct.stroke();
    }

    function pathLine()
    {
      ct.beginPath();
      ct.moveTo(225, 25);
      ct.lineTo(300, 50);
      ct.lineTo(300, 100);
      ct.lineTo(225, 125);
      ct.fill();
      ct.stroke();
    }
...
```

Listing 15.5 canvas.htm File, Upper Part

In the rectangle() function, you use the fillRect() method to create a filled rectangle using the currently set fill type and color. The first two parameters represent the *x* and *y* coordinates of the top left-hand corner, and they indicate the distance of this corner from the left or top edge of the canvas. The next two parameters specify the values for the width and height of the rectangle.

The strokeRect() method creates an empty rectangle with the currently set line type and color. The same parameters apply, so the empty rectangle forms the frame for the filled rectangle.

You create filled or empty arcs or filled or empty polylines along a path. In the pathArc() function, you initialize the path with the beginPath() method.

The arc() method creates a circular arc. The first two parameters represent the *x* and *y* coordinates of the center around which you draw the arc, and the next parameter specifies the radius of the arc.

The last two parameters indicate the start angle and the end angle. The angle is given in radians, and in the example, the start angle is 0 degrees. As usual, this is at 03:00 a.m., and it goes clockwise to an angle of 270 degrees (i.e., 12:00 p.m.). A radian value of 2 π corresponds to an angle of 360 degrees.

The fill() method creates the outline of a path and fills it, and you draw the fill between the start and end points of the path. The stroke() method only draws the outline of a path.

The path in the pathLine() function consists of individual line segments, while the moveTo() method moves the drawing pencil to the point with the specified *x* and *y* coordinates. The lineTo() method draws a line as a connection to the specified point.

Now, here's the middle part of the document with the functions for creating gradients and deleting the canvas:

```
...
    function gLinear()
    {
        const gr = ct.createLinearGradient(5, 150, 80, 250);
        gr.addColorStop(0.0, "#000000");
        gr.addColorStop(0.5, "#ffffff");
        gr.addColorStop(1.0, "#000000");
        ct.fillStyle = gr;

        ct.fillRect(5, 150, 75, 100);
        ct.fillStyle = "#f0f0f0";
    }

    function gRadial()
    {
```

15

```
        const gr = ct.createRadialGradient(150, 200, 0, 150, 200, 50);
        gr.addColorStop(0.0, "#000000");
        gr.addColorStop(0.5, "#ffffff");
        gr.addColorStop(1.0, "#000000");
        ct.fillStyle = gr;

        ct.beginPath();
        ct.arc(150, 200, 50, 0, 2 * Math.PI);
        ct.fill();

        ct.fillStyle = "#f0f0f0";
      }

    function clear()
    {
        ct.clearRect(0, 0, 400, 300);
    }
  </script>
</head>
...
```

Listing 15.6 canvas.htm File, Middle Part

You generate a linear gradient in the gLinear() function. You must imagine the gradient as a straight line, and the fill color of a graphic object changes along this line. The first two parameters of the createLinearGradient() method represent the x and y coordinates of the starting point of the line, and the next two parameters represent the x and y coordinates of the end point of the line.

The method returns an object that provides access to the line. The addColorStop() method assigns specific colors to individual points on the line, and the first parameter identifies the point on the line. The 0.0 point stands for the starting point of the line, the 1.0 point stands for the end point of the line, and all values between 0 and 1 are in between (for example, the 0.5 point is at the halfway point).

Once you create the gradient, it gets assigned to the fillStyle property as the current fill type and color. To illustrate the color gradient, you can create a filled rectangle that uses the gradient, and you can select the coordinates of the gradient so that the top left corner of the rectangle is at the starting point of the gradient and the bottom right corner is at the end point. You can then set a different fill type and color for graphic objects you subsequently create.

You create a radial color gradient in the gRadial() function. The createRadialGradient() method expects information on two circles: the first circle marks the start of the color gradient, and the second circle marks its end. You create the circles one after the other using the *x* and *y* coordinates of the center and the radius. This results in the six parameters of the method.

To illustrate the color gradient, you can create an arc that uses the gradient. The arc is complete from 0 degrees to 360 degrees, and you select the coordinates of the gradient so that the center of the arc lies on the first circle and the edge of the arc lies on the second circle.

15.2.2   Images

You can use the drawImage() method to display an image in a canvas. You can add the image to the drawing in its original size or scaled, or you can add a section of the image (see the example in Figure 15.4).

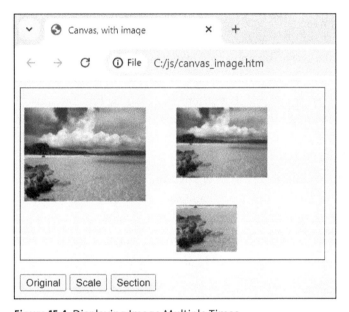

Figure 15.4 Displaying Image Multiple Times

Here's the program code:

```
...
<body>
   <canvas id="idCanvas" width="400" height="220"
         style="border:solid 1px #000000;"></canvas>
   <p><input id="idOriginal" type="button" value="Original">
      <input id="idScale" type="button" value="Scale">
      <input id="idSection" type="button" value="Section"></p>
   <script>
      const c = document.getElementById("idCanvas");
      const ct = c.getContext("2d");
```

```
        const image = document.createElement("img");
        image.src = "im_paradise.jpg";

        document.getElementById("idOriginal").addEventListener
            ("click", function() {ct.drawImage(image, 5, 25);} );
        document.getElementById("idScale").addEventListener
            ("click", function()
            {ct.drawImage(image, 205, 25, 120, 90);} );
        document.getElementById("idSection").addEventListener
            ("click", function()
            {ct.drawImage(image, 0, 60, 80, 60, 205, 150, 80, 60);} );
    </script>
</body></html>
```

Listing 15.7 canvas_image.htm File

The canvas has a size of 400 × 220 pixels and a thin black frame. Three buttons for creating the three objects follow.

In the JavaScript section, you create a new HTML element of the img type by using the createElement() method, and a reference to it is assigned to the image variable. The name of the image file is assigned to the property src of the HTML element, but the image is not yet added to the document.

You can display the HTML element in the document by using the drawImage() method. Three event handlers follow, and they lead to different calls of the method. You can call the method with three, five, or nine parameters that do the following:

- The first parameter references an HTML element of the img type that gets displayed.
- Parameters 2 and 3 represent the x and y coordinates of the top left-hand corner of the image.
- Parameters 4 and 5 specify the visible width and height of the image. The aspect ratio should be maintained.
- If nine parameters are passed during the call, parameters 2 to 5 are moved backwards. The new parameters 2 and 3 mark the top left-hand corner of the image section within the image, and the size of the image section follows in parameters 4 and 5. As with parameters 2 to 5 of the method call before, parameters 6 to 9 represent the coordinates and the visible size of the image.

15.2.3   Formatted Texts

The fillText() and strokeText() methods create a filled or outlined text within a drawing, and you use the textBaseline property to define the baseline on which the text is written (see the example in Figure 15.5).

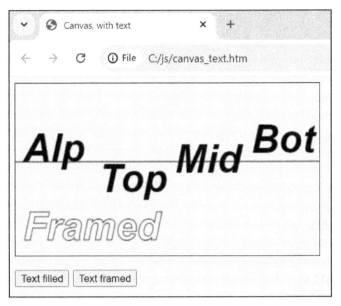

Figure 15.5 Integrating Text

Let's first look at the lower part of the document:

```
...
<body>
    <canvas id="idCanvas" width="400" height="220"
            style="border:solid 1px #000000;"></canvas>
    <p><input id="idFilled" type="button" value="Text filled">
        <input id="idFramed" type="button" value="Text framed"></p>
    <script>
        const c = document.getElementById("idCanvas");
        const ct = c.getContext("2d");
        ct.fillStyle = "#000000";
        ct.lineWidth = 1;
        ct.strokeStyle = "#000000";
        ct.font = "italic bold 50px Arial";

        document.getElementById("idFilled").addEventListener("click", filled);
        document.getElementById("idFramed").addEventListener("click", framed);
    </script>
</body></html>
```

Listing 15.8 canvas_text.htm File, Lower Part

At the beginning, some default settings are made for the lines and the fills, as you have already seen with arcs, for example. You use the font property to define the font properties.

Now, here's the upper part of the document:

```
... <head> ...
   <script>
      function filled()
      {
          ct.beginPath();
          ct.moveTo(0, 100);
          ct.lineTo(400, 100);
          ct.stroke();

          ct.textBaseline = "alphabetic";
          ct.fillText("Alp", 10, 100);
          ct.textBaseline = "top";
          ct.fillText("Top", 110, 100);
          ct.textBaseline = "middle";
          ct.fillText("Mid", 210, 100);
          ct.textBaseline = "bottom";
          ct.fillText("Bot", 310, 100);
      }

      function framed()
      {
          ct.textBaseline = "alphabetic";
          ct.strokeText("Framed", 10, 200);
      }
   </script>
</head>
...
```

Listing 15.9 canvas_text.htm File, Upper Part

In the filled() function, a ledger line is first drawn to clarify the text baseline. Different values are then used for the different distances to the text baseline, and an inscription is generated in each case. The first parameter of the fillText() method contains the output text, and the next two parameters contain the position of the bottom left corner of the first character.

The default alphabetic value causes most characters to be exactly on the line. Only the lowercase letters g, j, p, q, and y protrude downward beyond the line. With the top value, the line runs above the inscription, with middle through the middle and bottom below it.

In the framed() function, you use the strokeText() method to create a framed text. The parameters are the same as for the fillText() method.

15.3   Sensors

Modern browsers in mobile devices provide the option of accessing data from the device sensors for geolocation, position, and acceleration. The values determined can be displayed using JavaScript and used for control purposes.

15.3.1   Geolocation

You can use the `geolocation` property of the `navigator` object to access geolocation data, and you can use this data as figures or on a map.

In the following *geolocation.htm* program, a timestamp, the latitude, and the longitude of the geolocation are each output in two ways (see Figure 15.6). After you call a program that requests your geolocation data, the program will first ask you for your consent.

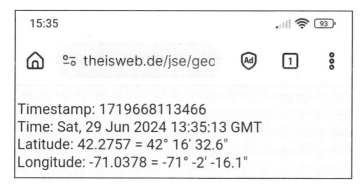

Figure 15.6 Geolocation Data

> **Note**
> You may have to call a sensor program several times before the sensor reacts and supplies data for the first time.

You can enter the decimal values for latitude and longitude in the Google Maps search field, for example, separating them with a space. The geolocation will then be displayed on the map.

Let's first look at the lower part of the program:

```
...
<body>
    <div id="idInfo"> </div>
    <script>
        const info = document.getElementById("idInfo");
```

```
      if(navigator.geolocation)
         navigator.geolocation.getCurrentPosition(receive, error);
      else
         error();
   </script>
</body></html>
```

Listing 15.10 geolocation.htm File, Lower Part

You use the `div` area with the `idInfo` ID to output the geolocation data. The `info` variable references this `div` area.

The next step is to check whether the browser recognizes the `geolocation` property. In the standard case, you are also asked whether the geolocation data may be retrieved during use. If the browser recognizes the property and you give consent to retrieve the geolocation data, the geolocation data is requested using the `getCurrentPosition()` method. Two references to functions are passed as parameters.

The successfully received data is output in the `receive()` function. If the browser doesn't know the property or if you deny consent to retrieve the geolocation data, the `error()` function throws an error message.

Here's the upper part of the program:

```
...  <head> ...
   <script>
      function receive(e)
      {
         const lat = e.coords.latitude;
         const latInteger = parseInt(lat);
         const latDecimalPlace = (lat - latInteger) * 60;
         const latMinutes = parseInt(latDecimalPlace);
         const latSeconds
            = (latDecimalPlace - parseInt(latDecimalPlace)) * 60;

         const lng = e.coords.longitude;
         const lngInteger = parseInt(lng);
         const lngDecimalPlace = (lng - lngInteger) * 60;
         const lngMinutes = parseInt(lngDecimalPlace);
         const lngSeconds
            = (lngDecimalPlace - parseInt(lngDecimalPlace)) * 60;

         const time = new Date(e.timestamp);
         info.innerHTML = "<p>Timestamp: " + e.timestamp + "<br>"
            + "Time: " + time.toUTCString() + "<br>"
```

```
                + "Latitude: " + lat.toFixed(4)
                + " = " + latInteger + "&deg; "
                + latMinutes + "&prime; "
                + latSeconds.toFixed(1) + "&Prime;<br>"
                + "Longitude: " + lng.toFixed(4)
                + " = " + lngInteger + "&deg; "
                + lngMinutes + "&prime; "
                + lngSeconds.toFixed(1) + "&Prime;</p>";
        }

        function error()
        {
            info.innerHTML = "No positioning";
        }
    </script>
</head>
...
```

Listing 15.11 geolocation.htm File, Upper Part

An object with the geolocation data is passed to the receive() function. This object contains the timestamp and coords properties. The timestamp can be converted into a Date object, and the coords property includes the latitude and longitude properties for the latitude and longitude. These values are output, rounded to four decimal places.

The sample values in Figure 15.6 are positive for the latitude and negative for the longitude. It's a northern latitude (north of the equator) and a western longitude (west of Greenwich, UK). The degree values have decimal places, which are converted here for the usual representation: in degrees, arc minutes, and arc seconds. The latitude with the value 0 degrees lies on the equator, the North Pole lies at a latitude of 90 degrees, and the South Pole lies at a latitude of –90 degrees. The longitude with the value 0 degrees is also called the *prime meridian*, and it runs through Greenwich and, like all longitudes, through the two poles. The longitude with the value of 180 degrees or –180 degrees runs near New Zealand, among other places, and it represents the international date line.

We explain the conversion to arc minutes and arc seconds below for the sample value of 42.2757 degrees from Figure 15.6. The decimal value (dv) results in the following: Auxiliary value $x = dv * 60$, bm = integer part of x, bs = (decimal places of x) * 60. With dv = 0.2757 follows: $x = dv * 60 = 16.542$, $bm = 16$, $bs = 0.542 * 60 = 32.52$, rounded to one decimal place = 32.5 (i.e., 42°16'32.5"). Conversely, the decimal places for a decimal degree can be calculated as follows: ($bm * 60 + bs$) / 3600. The result for this example is (16 * 60 + 32.52) / 3600 = 0.2757.

15

423

15.3.2 Waytracking

You can use the collection of geolocation data to determine the temporal and geographical course of a route you take on foot, by bike, or in a car. In the downloadable material for this book, you'll find the *waytracking.htm* program as a bonus with many explanatory comments.

In addition to the data you already know from Section 15.3.1, you can use the `coords.altitude` subproperty to record the altitude above sea level. The program also takes into account the possibility that this value may not be recorded in the mobile device.

After you click the **Start** button, the geolocation data is recorded every five seconds and added as an additional row to the table. When you click the **Stop** button, the recording and output will end. You can see the header of the table in Figure 15.7.

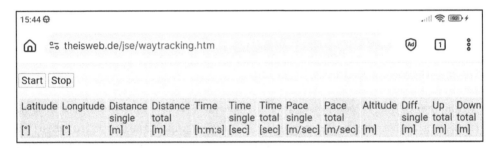

Figure 15.7 Waytracking Table Header

Particularly at low speeds, with low altitude differences, and with frequent detection, the values determined may deviate slightly from the actual values. Experience has shown that the deviations even out as the distance increases.

Here's a description of the table columns:

- *Latitude* in degrees: `coords.latitude` property
- *Longitude* in degrees: `coords.longitude` property
- *Single distance* in meters: Distance covered on an orthodrome (see below) between the current detection point and the previous detection point
- *Total distance* in meters: Sum of the individual paths
- *Time* in hours, minutes, and seconds: `timestamp` property
- *Single time* in seconds: Time difference between the current detection point and the previous detection point
- *Total time* in seconds: Time difference between the current detection point and the first detection point
- *Single speed* in meters per second: Single distance divided by single time
- *Total speed* in meters per second: Total distance divided by total time

- *Altitude* in meters: `coords.altitude` property
- *Single altitude difference* in meters: Difference in altitude between the current detection point and the previous detection point
- *Total ascent* in meters: Total of the individual altitude differences, if it's a positive difference of at least one meter
- *Total descent* in meters: Total of the individual altitude differences, if it's a negative difference of at least one meter

An orthodrome is the shortest path between two points on the surface of a sphere. You can find the source for the conversion at *https://en.wikipedia.org/wiki/Great-circle_distance.*

An average value of approximately five seconds appears for single time. The individual values fluctuate, especially at the beginning, as there are different time differences between the attempt to record the data and the actual receipt of the data.

The total speed stabilizes after some time at a constant speed. Due to the inaccuracies in the recording, I decided to only take into account altitude differences of more than one meter for the calculation of total ascent and total descent.

Note

Make sure that the display of your mobile device doesn't switch off before the end of the automatic detection because you are no longer operating the mobile device. That could cause no more data to be recorded.

15.3.3 Position Sensor

You can use the `DeviceOrientationEvent` object, which processes the `deviceorientation` event. This allows a change in orientation to be detected and received by the position sensors in a mobile device. Here's an example of how the position sensors work: if you rotate your mobile device, the display will switch between portrait format and landscape format and back again.

You can then output the changed position data, but you can also process them further. You can use this for games and simulations, for example. In the following *orientation.htm* program, you can see a simple further processing for controlling an object on the screen (see Figure 15.8).

The position information comprises three values: the `alpha`, `beta`, and `gamma` angles. The `alpha` angle indicates the angle between the longitudinal axis of the mobile device and the north pole (see Figure 15.9). Values between 0 and 360 degrees are possible here.

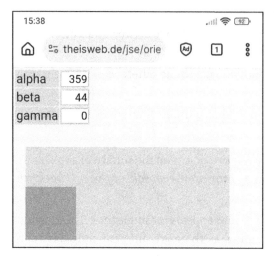

Figure 15.8 Position Data, Output, and Control

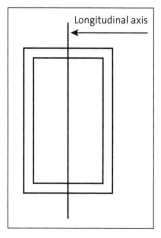

Figure 15.9 Longitudinal Axis of Mobile Device

To illustrate these angles, it's best to first lay your mobile device flat on the table in front of you. Then do the following:

- If you turn the mobile device while it's still lying flat on the table, the alpha angle changes. At 0 degrees or 360 degrees, the longitudinal axis of the mobile device points to the north pole.

- If you pull up the top edge of the mobile device until it's standing upright in front of you, you'll notice that the beta angle slowly changes from 0 degrees to 90 degrees. Do the same with the bottom edge of the mobile device (i.e., tilt it away from you until it's completely upside down), and the beta angle changes from 0 degrees to −90 degrees. Overall, values from −180 degrees to +180 degrees are possible for this angle.

- If you lift up the left edge and tilt the mobile device to the side, the gamma angle changes from 0 degrees to 90 degrees, while raising the right edge changes the gamma angle from 0 degrees to –90 degrees. Here, too, total values from –180 degrees to +180 degrees are possible.

In addition, the dark square in the *orientation.htm* program moves inside the light rectangle when the device is tilted. Thus, the change in position data is used to move an object.

Let's look at the lower part of the program:

```
...
<body>
   <table>
      <tr><td>alpha</td><td><input id="idAlpha"
         style="text-align:right; width:100px;" readonly></td></tr>
      <tr><td>beta</td> <td><input id="idBeta"
         style="text-align:right; width:100px;" readonly></td></tr>
      <tr><td>gamma</td><td><input id="idGamma"
         style="text-align:right; width:100px;" readonly></td></tr>
   </table>
   <div style="position:absolute; top:300px; left:50px;
      width:800px; height:350px; background-color:#e0e0e0;"></div>
   <div id="idRect" style="position:absolute; top:300px; left:200px;
      width:200px; height:200px; background-color:#b0b0b0;"></div>
   <script>
      const rect = document.getElementById("idRect");
      if(DeviceOrientationEvent)
         addEventListener("deviceorientation", position, false);
   </script>
</body></html>
```

Listing 15.12 orientation.htm File, Lower Part

There's a table containing three read-only input fields for outputting the position data, and there are div areas for the dark square and the light rectangle positioned below the table.

The system checks whether the browser recognizes the DeviceOrientationEvent object. If it does, then an event listener for the deviceorientation event will get registered for the window object. The position() function is therefore called when the position changes.

Here's the upper part of the program:

```
... <head> ...
   <script>
```

```
    let moveReference, pTop = 300, pLeft = 200;

    function position(e)
    {
        clearTimeout(moveReference);
        document.getElementById("idAlpha").value = Math.round(e.alpha);
        document.getElementById("idBeta").value = Math.round(e.beta);
        document.getElementById("idGamma").value = Math.round(e.gamma);
        move(e.beta, e.gamma);
    }

    function move(b,g)
    {
        if(b > 3 && pTop < 450)        pTop += 5;
        else if(b < -3 && pTop > 300) pTop -= 5;
        rect.style.top = pTop + "px";

        if(g > 3 && pLeft < 650)        pLeft += 5;
        else if(g < -3 && pLeft > 50) pLeft -= 5;
        rect.style.left = pLeft + "px";

        moveReference = setTimeout(function() { move(b,g); }, 50);
    }
    </script>
</head>
...
```

Listing 15.13 orientation.htm File, Upper Part

To change the position of the dark square, it's first necessary to store the initial values for the position in the pTop and pLeft variables.

In addition, a variable is introduced as a reference to a time-controlled process. What is this reference needed for?

The deviceorientation event only occurs when the position changes, so if you tilt your mobile device and then hold it still in the tilted position, the event will no longer occur. However, the dark square should continue to move in accordance with the laws of physics. For this reason, every time the position changes, a time-controlled process starts and continues the movement continuously until a new position change occurs. Only the new position change is then evaluated in the same way.

An object with the position data is transferred to the position() function, and this object contains the alpha, beta, and gamma properties. Then the timer for the current movement gets deleted, the move() function is called, and the beta and gamma angles are transferred to the move() function.

The move() function checks whether the device is tilted by at least three degrees in one of the four possible directions. This compensates for small wobbles of your hand, and you have the chance to stop the movement of the dark square by holding your mobile device straight. Secondly, the function checks whether the dark square has already reached the edge of the light rectangle. It should not move beyond the edge.

If the checks are successful, the dark square gets moved by five pixels. The function then calls itself again fifty milliseconds later with the same position data so that the movement continues smoothly. If the mobile device is given a different position in the meantime, the time-controlled process in the position() function terminates and the new position change gets initiated.

15.3.4 Acceleration Sensor

You can use the DeviceMotionEvent object, which processes the devicemotion event. This allows the *acceleration* (i.e., the change in speed) to be measured.

You can output this acceleration data and process it further (e.g., for a game or a simulation). In the following *motion.htm* program, an object is controlled on the screen (see Figure 15.10), but different movements of the mobile device are required than for processing the position data.

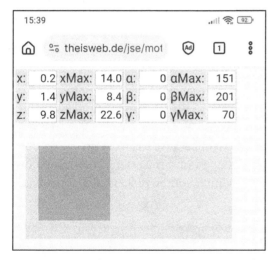

Figure 15.10 Acceleration Data, Maximums, and Control

A Note on Physics

If an object moves in a certain direction, it has a speed that is measured in meters per second. If this speed changes (for instance, when a vehicle starts or brakes), this is called acceleration, which corresponds to the change in speed per unit of time and is measured in meters per second squared (which we write as m/s^2).

If you move your mobile device at a constant speed in a certain direction, there will be no acceleration. The sensors only measure acceleration and trigger the `devicemotion` event when the mobile device is brought to a standstill or when movement is initiated.

All objects on earth are constantly accelerated towards the center of the earth due to *gravity*. For simplicity's sake, this acceleration is also referred to as gravitational acceleration, the standard value of which at the earth's surface is approximately 9.81 m/s^2 = 1 g. This explains the value for z.

In Figure 15.10, you can see values for x, y, z, `alpha`, `beta`, and `gamma`. To clarify these values, it's best to first place your mobile device flat in front of you on a table with a smooth surface and then do the following:

- Move your device rapidly to the right on the table and slow it down until it stops. A positive x value is measured briefly at the beginning of the process, and a negative x value is measured briefly at the end.

- Move your device rapidly away from you on the table and slow it down until it stops. A positive y value is measured briefly at the beginning of the process, and a negative y value is measured briefly at the end.

- Pick up your device, lift it toward the ceiling rapidly, and slow it down until it stops. At the start of the sequence, a z value greater than 9.81 is measured for a short time, as the acceleration you have applied is added to the gravitational acceleration. At the end, a z value of less than 9.81 is measured for a short time.

These accelerations are used to change the position and size of the dark square in the light rectangle.

In the x and y directions, the square shows sluggish behavior due to the program. When you move your mobile device rapidly to the right on the table and slow it down until it stops, the square initially moves to the left because it seemingly can't follow the acceleration as quickly. At the end, it moves to the right because it can't seem to slow down as quickly. When you lift your device rapidly up toward the ceiling and slow it down until it stops, the square is smaller at the beginning of the process, and at the end, it gets bigger again.

Another Note on Physics

The speed and acceleration of an object in a certain direction are also referred to as *orbital speed* and *orbital acceleration* because the object moves along an orbit. In contrast, there's angular velocity and angular acceleration, which are measured in degrees per second and degrees per second squared, respectively.

Mobile devices can measure angular acceleration. To clarify this parameter, lay your device flat on the table in front of you. Rotating the device while it's still lying on the table leads to an acceleration by the `gamma` angle, raising the top edge of the device quickly leads to an acceleration by the alpha angle, and raising the side edge of the device quickly leads to an acceleration by the beta angle.

All six of these accelerations only occur for a short time, and for this reason, the maximum values are permanently displayed in the program.

Here's the lower part of the document:

```
...
<body>
    <table>
        <tr>
            <td>x:</td><td><input id="idX"
                style="text-align:right; width:100px;" readonly></td>
            <td>xMax:</td><td><input id="idXMax"
                style="text-align:right; width:100px;" readonly></td>
            <td>&alpha;:</td><td><input id="idAlpha"
                style="text-align:right; width:100px;" readonly></td>
            <td>&alpha;Max:</td><td><input id="idAlphaMax"
                style="text-align:right; width:100px;" readonly></td>
        </tr>
        <tr>
            <td>y:</td> <td><input id="idY"
                style="text-align:right; width:100px;" readonly></td>
            <td>yMax:</td><td><input id="idYMax"
                style="text-align:right; width:100px;" readonly></td>
            <td>&beta;:</td> <td><input id="idBeta"
                style="text-align:right; width:100px;" readonly></td>
            <td>&beta;Max:</td> <td><input id="idBetaMax"
                style="text-align:right; width:100px;" readonly></td>
        </tr>
        <tr>
            <td>z:</td><td><input id="idZ"
                style="text-align:right; width:100px;" readonly></td>
            <td>zMax:</td><td><input id="idZMax"
                style="text-align:right; width:100px;" readonly></td>
            <td>&gamma;:</td><td><input id="idGamma"
                style="text-align:right; width:100px;" readonly></td>
            <td>&gamma;Max:</td><td><input id="idGammaMax"
                style="text-align:right; width:100px;" readonly></td>
        </tr>
    </table>
    <div style="position:absolute; top:300px; left:50px;
        width:800px; height:350px; background-color:#e0e0e0;"></div>
    <div id="idRect" style="position:absolute; top:300px; left:200px;
        width:200px; height:200px; background-color:#b0b0b0;"></div>
```

15

```
<script>
   const rect = document.getElementById("idRect");
   if(DeviceMotionEvent)
      addEventListener("devicemotion", acceleration, false);
</script>
</body></html>
```

Listing 15.14 motion.htm File, Lower Part

There are a total of twelve write-protected input fields in the table for outputting the six measured values and the six associated maximum values. This is followed by the two div areas for the dark square and the light rectangle.

The system checks whether the browser recognizes the DeviceMotionEvent object. If it does, then an event listener for the devicemotion event is registered for the window object. This means that the acceleration() function is called during acceleration.

Here's the upper part of the program:

```
... <head> ...
   <script>
      let pTop = 300, pLeft = 200, size = 200;
      let xMax = 0, yMax = 0, zMax = 9.81;
      let alphaMax = 0, betaMax = 0, gammaMax = 0;

      function acceleration(e)
      {
         const acc = e.accelerationIncludingGravity;

         document.getElementById("idX").value = acc.x.toFixed(1);
         document.getElementById("idY").value = acc.y.toFixed(1);
         document.getElementById("idZ").value = acc.z.toFixed(1);

         if(Math.abs(acc.x) > xMax)
         {
            xMax = Math.abs(acc.x);
            document.getElementById("idXMax").value = xMax.toFixed(1);
         }
         if(Math.abs(acc.y) > yMax)
         {
            yMax = Math.abs(acc.y);
            document.getElementById("idYMax").value = yMax.toFixed(1);
         }
         if(Math.abs(acc.z) > zMax)
         {
```

```
        zMax = Math.abs(acc.z);
        document.getElementById("idZMax").value = zMax.toFixed(1);
    }

    if(acc.x > 3 && pLeft > 50)   pLeft -= 50;
    if(acc.x < -3 && pLeft < 650) pLeft += 50;
    rect.style.left = pLeft + "px";

    if(acc.y < -3 && pTop > 300) pTop -= 50;
    if(acc.y > 3 && pTop < 450)  pTop += 50;
    rect.style.top = pTop + "px";

    if(acc.z > 12.81 && size > 100) size -= 20;
    if(acc.z < 6.81 && size < 300)  size += 20;
    rect.style.width = size + "px";
    rect.style.height = size + "px";

    if(e.rotationRate)
    {
        const rot = e.rotationRate;
        document.getElementById("idAlpha").value = Math.round(rot.alpha);
        document.getElementById("idBeta").value = Math.round(rot.beta);
        document.getElementById("idGamma").value = Math.round(rot.gamma);

        if(Math.abs(rot.alpha) > alphaMax)
        {
            alphaMax = Math.abs(rot.alpha);
            document.getElementById("idAlphaMax").value = Math.round(alphaMax);
        }
        if(Math.abs(rot.beta) > betaMax)
        {
            betaMax = Math.abs(rot.beta);
            document.getElementById("idBetaMax").value = Math.round(betaMax);
        }
        if(Math.abs(rot.gamma) > gammaMax)
        {
            gammaMax = Math.abs(rot.gamma);
            document.getElementById("idGammaMax").value = Math.round(gammaMax);
        }          }
    }
    </script>
</head>
...
```

Listing 15.15 motion.htm File, Upper Part

The initial values for the position and size of the dark square are recorded in the pTop, pLeft, and size variables. In addition, the six maximum values are set to 0, with one exception: the maximum value for z is given the starting value of 9.81.

An object with the acceleration data is passed to the acceleration() function, and among other things, this object has the acceleration, accelerationIncludingGravity, and rotationRate properties. The first two of these properties contain the x, y, and z properties for the different orbital accelerations. The difference is that if the mobile device is at rest, z has the value 0 for the acceleration object; otherwise, it has the value 9.81.

The program then checks whether one of the acceleration values deviates by at least 3 m/s^2 from the value that prevails in the rest position, and it also checks whether the dark square has reached the edge of the light rectangle. The square should not move beyond the edge.

If the checks have a positive result, the dark square gets moved by fifty pixels or becomes twenty pixels larger or smaller.

On some mobile devices, the angular accelerations can't be output, and the corresponding rotationRate object is therefore queried beforehand. This object contains the three different values in the alpha, beta, and gamma properties.

For all six acceleration values, the system checks whether the current value is greater than the value of the relevant maximum. If it is greater, then the new maximum will be saved and output.

Operating the program correctly (i.e., moving the square in a specific direction) is not easy, so you should initiate the movement slowly and end it abruptly. An acceleration of more than 3 m/s^2 only occurs during deceleration, and that makes the effect easier to see.

One last tip: hold the mobile device firmly when testing the accelerations.

Appendix A
Installation and Keywords

This appendix describes the installation of *XAMPP* and then provides a list of keywords in the JavaScript language. For the use of XAMPP, please refer to Chapter 4, Section 4.2.2.

A.1 Installation of the XAMPP Package

In this appendix, I describe the installation of the free *XAMPP* package, which you can use to test PHP programs. The currently available version of XAMPP (as of August 2024) is version 8.2.12.

A.1.1 Installation on Windows

You can download the XAMPP for Windows package from *https://www.apache-friends.org*.

Then, start the installation by calling the executable file, *xampp-windows-x64-8.2.12-0-VS16-installer.exe*. Two warnings may appear at the beginning, indicating that an anti-virus program is running, among other things. You can continue here and confirm the suggested installation options. I recommend selecting *C:\xampp* as the target directory.

After the installation, start the XAMPP Control Panel application, where you can start and stop the Apache web server using the button to the right of the word *Apache*.

If you receive the Unable to open process error message when starting the server, it may be due to a port that is already being used. You can find clear instructions on how to rectify this problem at the following address: *http://www.coder-welten.com/windows-10-unable-to-open-process*. (Note that this very descriptive article is in German. In a modern browser, you can have it automatically translated.)

You can reach the start page of the local web server in your browser via the *localhost* address, and you can save your HTML files and PHP programs in the *C:\xampp\htdocs* directory and in the directories below it.

The following PHP program enables you to check whether the installation was successful. You can write it using an editor and save it in the *C:\xampp\htdocs\phpinfo.php* file:

```php
<?php
    phpinfo();
?>
```

Listing A.1 phpinfo.php File

After you enter the corresponding *localhost/phpinfo.php* address in the browser, information about the installed PHP version appears.

Stop the Apache web server in the XAMPP Control Panel application, then close the application.

A.1.2   Installation on Ubuntu Linux

You can download the XAMPP for Linux package from *https://www.apachefriends.org/*. After the download, the *xampp-linux-x64-8.2.12-0-installer.run* file is available.

Open a terminal to enter the installation commands, and if necessary, change the access rights to the file by using the following command:

```
chmod 744 xampp-linux-x64-8.2.12-0-installer.run
```

Start the installation via this command:

```
sudo ./xampp-linux-x64-8.2.12-0-installer.run
```

You can confirm the suggested installation options, and XAMPP will be installed in the */opt/lampp* directory.

At the end of the installation, you can leave the **Launch Xampp** checkbox checked. This opens a dialog box for managing the various servers. On the **Manage Servers** tab, you have the option of selecting the Apache web server and starting and stopping it using the button on the right. You can also call the dialog box for managing the servers directly as follows:

```
sudo /opt/lampp/manager-linux-x64.run
```

However, the Apache web server of Ubuntu Linux is often already running after a system start. If it is running, the Apache web server of XAMPP can't start. If that happens, you must first terminate the Apache web server of Ubuntu Linux once, using the following command:

```
sudo /etc/init.d/apache2 stop
```

After that, you can start the Apache web server of XAMPP as described previously.

To avoid having to close the Apache web server of Ubuntu Linux after every system start, you can deactivate the automatic start of the associated service with the following command:

```
sudo update-rc.d apache2 disable
```

If you want to reactivate the automatic start of the associated service at some point, you can do so via the following command:

```
sudo update-rc.d apache2 enable
```

You can reach the start page of the local web server in your browser via the *localhost* address, and then, you can save your HTML files and PHP programs in the */opt/lampp/ htdocs* directory and in the directories below it.

To test whether the installation was successful, you can use the PHP program below. You can write and save it in the *phpinfo.php* file using the *nano* editor as follows:

```
sudo nano /opt/lampp/htdocs/phpinfo.php
```

You can exit the nano editor as follows: press Ctrl+X, confirm with **Yes** at the bottom under **Save**, and confirm the suggested file name. The PHP program looks as follows:

```php
<?php
   phpinfo();
?>
```

Listing A.2 phpinfo.php File

After you enter the corresponding *localhost/phpinfo.php* address in the browser, information about the installed PHP version appears.

A.1.3 Installation on macOS

You can download the XAMPP for OS X package from *https://www.apachefriends.org*. After that, the *xampp-osx-8.2.12-0-installer.dmg* file is available. Double-click on this file to create a new drive, and then, you can call the installation file located on the new drive.

Since the installation program doesn't come from a verified Apple developer, it gets initially blocked by macOS. Call the **Security** menu item in the system settings, and there, you'll find the **Open anyway** button next to the reference to XAMPP. After clicking that button, you'll have the option of starting the installation program.

By default, XAMPP is installed in the */Applications/XAMPP* directory, which corresponds to the *Applications/XAMPP* directory in the Finder.

At the end of the installation, you can leave the **Launch Xampp** checkbox checked. This opens a dialog box for managing the various servers. On the **Manage Servers** tab, you have the option of selecting the Apache web server and starting and stopping it using the button on the right. You can also call the dialog box for managing the various servers via *Applications/XAMPP/manager-osx*.

You can reach the start page of the local web server in your browser via the *localhost* address, and you can save your HTML files and PHP programs in the *Applications/ XAMPP/htdocs* directory and in the directories below it.

To test whether the installation was successful, you can use the following PHP program. You can write and save it in the *Applications/XAMPP/htdocs/phpinfo.php* file using the TextWrangler editor from the App Store.

The PHP program looks like this:

```
<?php
    phpinfo();
?>
```

Listing A.3 phpinfo.php File

After you enter the corresponding *localhost/phpinfo.php* address in the browser, information about the installed PHP version appears.

A.2   List of Keywords

When creating variables and your own functions, you should assign custom names that are as self-explanatory as possible. The assignment of names should follow certain rules (see Chapter 2, Section 2.1.2), and in particular, you may not use the following names of keywords of the JavaScript language because they have been or will be used by JavaScript itself:

await, abstract, boolean, break, byte, case, catch, char, class, const, continue, debugger, default, delete, do, double, else, enum, export, extends, false, final, finally, float, for, function, get, goto, if, implements, import, in, instanceof, int, interface, let, long, native, new, null, package, protected, private, prototype, public, return, set, short, super, static, switch, synchronized, this, throw, throws, true, transient, try, typeof, undefined, var, void, volatile, while, with, **and** yield.

Appendix B
The Author

 **Thomas Theis** has more than 40 years of experience as a software developer and as an IT lecturer. He holds a graduate degree in computer engineering. He has taught at numerous institutions, including the Aachen University of Applied Sciences. He is the author of several successful technical books on the topics of Python, C#, PHP, C++, and Unity.

Index